MADISON

A Classic Southern Town

William R. Mitchell Jr.

Photography by Van Jones Martin and James R. Lockhart

Foreword by Philip Lee Williams

A GOLDEN COAST BOOK

Produced for

THE HISTORIC MADISON-MORGAN FOUNDATION

Madison, Georgia

© 2009 Historic Madison-Morgan Foundation

Text © William R. Mitchell Jr.

All rights reserved. No part of this book may be reproduced in any form or by any electronic or mechanical means including information storage and retrieval systems without permission in writing from Historic Madison-Morgan Foundation, except by a reviewer, who may quote brief passages in a review.

Designed and produced for Historic Madison-Morgan Foundation by Golden Coast Publishing Company, Savannah, Georgia,
Van Jones Martin and Jo Morrell.
www.goldencoastbooks.com

Edited by Jane Powers Weldon, Marshall "Woody" Williams, Monica Callahan, and Van Jones Martin.

Digital production by David J. Kaminsky, Savannah Color Services.

ILLUSTRATION CREDITS:

Art Work of North Central Georgia, private collection: 43, bottom; 46, top left; 51, center; 52, top left; 53, top center, top right, and middle; 54, bottom; 55, top.

Georgia Archives: x, bottom; 34; 35, bottom; 36; 38–39; 44, top; 47, center; 49, center; 50, top; 52, top center; 55, bottom left, bottom center, second from top; 56, top; 57, center; 58, bottom.

Library of Congress: 12; 14; 15; 16–17, center; 45, bottom; 58, second from bottom.

© James R. Lockhart: 20, top; 22, bottom; 28, bottom; 60, all; 61, top, right; 65, bottom center and right; 66, top left and bottom; 76, bottom; 77; 94, top; 102–3, all; 106–7, all; 118–121, all; 123–27, all; 148, bottom; 149, bottom right; 150, bottom left; 158–59, all except 159, bottom right; 162–63, all except 163, bottom right; 164–66, all; 167, bottom left and right; 168–69, all; 171; 187, all; 188, bottom left and right; 189, bottom; 190, bottom; 191, top; 196, top; 232–33, all.

Van Jones Martin for the Georgia Department of Natural Resources, Historic Preservation Section: 41, bottom; 46, bottom; 47, top.

Morgan County Archives: x, top; 18–19; 35, top and middle; 42, bottom; 45, middle; 46, left column top; 49, bottom; 51, bottom; 52, second from bottom; 54, top left, second from bottom; 55, bottom right; 56 bottom; 58, top left; 59, top right; 61, center left.

Private Collections: xi; 17, top right; 23, bottom; 24, top and bottom; 43, top right; 44, center and bottom; 45, top; 47, bottom; 48; 51, top; 52, bottom; 53, top left and bottom; 54, top right; 57, top and bottom; 58, top right; 59, top left; 65, top.

Wikimedia Commons: 37; 41, top.

All photographs not otherwise attributed are © Van Jones Martin.

Printed in China by Everbest Printing Co., Ltd., through Four Colour Import Group.

ISBN 978-0-932958-27-3
Limited slipcase edition: ISBN 978-0-932958-28-1
Library of Congress Control Number: 2009911964

Half-title page: A scene on Academy Street, with Holly Hall (c. 1830s). Title page: Carter-Newton house (1850) on Academy.
Contents page: Sims-Speed-Bearden house (c. 1809) on East Washington Street. Acknowledgments page: East Washington Street.

Acknowledgments

This book was a collaborative effort and could not have been produced without the generous assistance of many civic-minded Madisonians and others in the county. We first wish to thank those who permitted us to visit their homes. Selecting from among them was a most difficult challenge. Every home visited would have made a worthy contribution to the book.
We offer special appreciation to Morgan County and the City of Madison, who helped in so many ways and whose bicentennial this book celebrates. Additionally, there is untold gratitude for the many friends from near and far who became founding members of the Historic Madison-Morgan Foundation so this book might become a reality.

Special acknowledgment must be extended to David and Shandon Land, whose original vision, diligent perseverance, and generous financial support launched and sustained this unique undertaking.

Bank of Madison, founded in 1890 and the county's oldest financial institution, deserves special appreciation for its extraordinary funding of the project.

For their faith in this celebratory endeavor, we sincerely thank those who gave so graciously:

Bank of Morgan County	Hall Smith Office, Inc.	Janet and Charlie Mason and the C. L. Mason Family
Bryans Family Foundation	Lowry and Lyn Hunt	Ginny and Dan Rather
In Memory of Virginia Conrads	Mrs. E. Roy (Chris) Lambert	Jane and Everett Royal
The Friesen Family	William A. and Maureen O. Lobb	Toby West and Tom Hayes

Others who were especially supportive of this book:

Sharon and Tim Barnes	Chandler Construction, Inc.	Madison-Morgan Conservancy	Rema Tip Top/North America, Inc.
Dan Belman and Randy Korando	James C. Conrads	Madison-Morgan Cultural Center	Sunshine Social Club
Ralph and Lucy Bennett	David and S. Alex DiRocco	Madison Town Committee, NSCDA-GA	Jane Campbell Symmes
Dean and Theresa Bishop	Bruce and Judy Ashurst Gilbert	Magnolia Garden Club	Robert and Anne Trulock
Donna and Ralph Blanchard	June Harrell	James J. McManus and Leticia Simback	Grady and Sally Tuell
Blue Stocking Book Club	Jimmy and Ellen Harrison	Mitzi and George Monroe	United Bank
The Bookies	Mary and John Huntz	Larry and Dell Morgan	Frank and Ann-Marie Walsh
Jim and Phyllis Boyd	The James Madison Inn	Morgan County *Citizen*	Watkins Builders, Inc.
Mrs. Frank (Jane) Carter Jr.	Ken and Monica Kocher	Morgan County Historical Society	RoseAnne and Russell Weaver
Wilson M. Carter	Robert Lanier and Dena Bordoni	Mitzi and Thomas Prochnow	Kathy and Clarence Whiteside

We extend our gratitude to Monica Callahan, City of Madison, and "Woody" Williams, former Morgan County Archivist, whose knowledge and assistance aided immeasurably in making this book what it is. The founding board of the book's publisher, the Historic Madison-Morgan Foundation, must be recognized for their work and oversight of the project: Cathy Best, Ruth Bracewell, June Harrell, Christine Lambert, David Land, Janet Mason, Adelaide Ponder, Mitzi Prochnow, Joseph Smith, Emmie Smock, and Jane Symmes.

Contents

Preface VIII
Foreword X
Introduction 12
The Town Sherman Refused to Burn—Fact or Fiction? 14
Town Plan, Early Architecture, and Landmarks 18
Education and the Madison-Morgan Cultural Center 23
Madison's Historic Churches and Congregations 26
Madisonians: People, Society, and Newsmakers 30
County and Region, Towns, Tourism, Hard Labor Creek State Park, and Lake Oconee 34
A Timeline of Historical Context for Madison and Morgan County 36

Madison Architecture and Preservation 62

Antebellum Architecture: 64
Log Cabins to Neoclassical
- Reuben Rogers House 68
- Edmund Walker Town House 70
- Richter Cottage 74
- Stagecoach House 76
- Cedar Lane Farm 78
- Hilltop 82
- Thurleston 88
- Barnett-Stokes House 94
- Robson-Mason House 98
- Bonar Hall 102
- Heritage Hall 104
- Martin-Baldwin-Weaver House 108
- Honeymoon 112
- Baldwin-Williford-Ruffin House 116
- Billups-Tuell House 122
- Massey-Tipton-Bracewell House 126
- Stokes-McHenry House 128
- Nathan Bennett House 134
- Boxwood 140

Postbellum Architecture: 148
Folk Victorian to Neoclassical Revival
- Rose Cottage 152
- Atkinson-Rhodes House 154
- Shaw-Erwin House 158
- Magnolia House 160
- La Flora 164
- White-Lyle Cottage 168
- Morgan County African-American Museum 170
- Godfrey-Hunt House 172
- Oak House 176
- Porter-Fitzpatrick House 182
- Poullain Heights 186

The Modern Era: 188
Styles and Trends, Historic Preservation, and Revitalization
- Bookhaven 192
- Carter-McManus House 194
- Douglas-McDowell House 196
- Burney-Ponder-Rushing House 198
- Gilbert House 202
- Samuel Hanson House 206
- Willow Oak Farm 212
- Stoke Farm 214
- Camp Boxwoods 216
- McFaddin Townhouse 224
- Hudson Loft 226
- Long Residence 230
- Royal Penthouse 232

Epilogue 234
Selected Sources and Observations 236
Index of Sites Illustrated 238

Preface

This commemorative Madison book is my eighteenth published volume and the twelfth book project on which Van Jones Martin and I have worked together. Our first collaboration, *Landmark Homes of Georgia, 1733–1983*, has a ten-page section about Madison/Morgan County in which we touched on the qualities that make this southern town an admired place to live and visit.

Van and I began this book because David Land of Madison, whom I had not met, called me in April 2008 to ask if I would be interested in helping him and his Madison bicentennial book committee do a volume to commemorate the town's two-hundredth anniversary in 2009. I said yes, but that we could use more time than was left at that point and suggested he call Van Martin to see if he would also be interested.

Van and I met with David for the first time May 1, 2008, with some other Madisonians at the Madison Chamber of Commerce. I came from Atlanta and Van from Savannah where he lives. One reason I could consider the project is because Madison is conveniently located sixty-five miles east of Atlanta, near I-20. Also, I was quite familiar with Madison; my own personal timeline with the place had begun when I was twenty-three, forty-eight years ago when I was a University of Delaware graduate student in American Studies shopping for a master's thesis subject.

I had arranged lunch with the noted antiquarian Henry D. Green (1909–2003) and traveled from Atlanta to his Morgan County home,

Cedar Lane Farm in 1982, from
Landmark Homes of Georgia, 1733–1983.

Madison is the prettiest village I've seen in the state. One garden and yard I never saw excelled, even in Connecticut.

Sgt. Rufus Mead, Jr.
November 19, 1864

Greenoaks Plantation, to discuss the restoration of the late eighteenth-century Redman Thornton house, a project that Green had chaired. I had watched as it progressed on the grounds of Atlanta's High Museum of Art on Peachtree Street in the late 1950s. After lunch Henry Green took me on a tour of Madison, including the Episcopal Church of the Advent renovation, for which he had been the church's senior warden and restoration advisor. I ended up using another topic for my thesis, but Henry Green, who was my father's age, was a fine choice as a tour guide for my introduction to Madison, nearly fifty years.

My Madison/Morgan timeline resumed in 1971, after I had begun my career and was head of the Georgia Historic Sites Survey, the state National Register program. I journeyed to the Morgan County restoration project of John and Jane Symmes, Cedar Lane Farm (the historic Hilsabeck house), to begin the process of putting it on the National Register of Historic Places, which was accomplished that year. Later in 1971, in July, I began working with Miss Therese Newton to put her family home, Bonar Hall (1832) located on Dixie Avenue, on the National Register, which we achieved in 1972.

In May 1974 the newly incorporated Georgia Trust for Historic Preservation held a regional preservation conference in Madison, Greensboro, and Eatonton. As a Georgia Trust board member I helped plan the conference and participated in the program, lecturing on the

Plantation Plain style of architecture. The meetings were held at the Rock Eagle 4-H Center, and there were tours and social events in the towns and countryside. We had events at Cedar Lane Farm and at Tom Cousins' Little River Farm (formerly the Henry Greens' Greenoaks, now owned by A. L. Williams). There were receptions and food at Hilltop, the Madison home of the E. Roy Lamberts on North Main Street, and at Elliott Acres (often referred to as the Rogers-Shields house), home of Mr. and Mrs. James Elliott Jr., next to Hilltop on North Main.

Almost a decade later in *Landmark Homes of Georgia*, we featured Boxwood, a major landmark of the Madison Historic District (added to the National Register in 1974), and Cedar Lane Farm in the county. I wrote: "These two houses are text-book examples of the two aspects of historic preservation: The Kolb-Newton house [Boxwood] is almost 'as is' inside and outside—a preservation in purest terms—and the Hilsabeck-Symmes house is one of the best examples of restoration and conservation to be found." I also featured a quote by Sergeant Rufus Mead Jr., a Connecticut Yankee with the Federal Army as it passed through Madison on its March to the Sea in November 1864. The quote, in which Mead called Madison the "prettiest village" he had seen in the state, has become part of the Madison story, and we have decided it bears visiting again as we try to capture the town's essence.

Over the years, I was in and out of Madison often, always dropping by the Cultural Center—opened in 1976—to see their latest exhibit, hear a lecture, or visit the center's excellent museum shop. On December 6, 1997, I attended one of Madison's well-known Christmas home tours, especially to see the house of Annie and Bob Jones, a postbellum house on North Main Street. (The Joneses, family friends, no longer own that house but are too fond of Madison to stay away for long and now live on Candler Street in a small enclave of "new urbanism." They could be said to represent a host of Madisonians who moved to town because it has always been the prettiest village.)

My own timeline culminated in this bicentennial book, which is a result of months of study and of new experiences with the setting, with its architecture, gardens, and landmarks, and the life of the place—its people, especially the gracious homeowners (and their cats and other pets). For me it has been a wonderful opportunity, coming towards the end of a long career that has included Madison/Morgan all along the way; the place is clearly an old friend.

Working again with Van Jones Martin, this project in the age of the internet, email, and digital photography has been a demanding but pleasurable exercise. I first knew Van during our Georgia Historic Site Survey days in 1972, when he was the state program's photographer before he started Golden Coast Publishing. Photographer Jim Lockhart, another old friend, has also been involved. Jim is the current state National Register photographer, and he and I have been associated on several book projects, including *J. Neel Reid, Architect*, published for the Georgia Trust in 1997. I've also made new Madison friends in David Land and his wife Shandon, who reside on Dixie Avenue in the historic Barnett-Stokes house, to which they have given new life, in the pattern of so many couples who have settled in Madison after seeing it on the way to somewhere else.

It has been a pleasure working with the Madison-Morgan Foundation, chaired by David Land, as it was being formed to publish this two-hundredth anniversary book, its first, but with the hope of future publications. Certainly, Madison/Morgan is a more than worthy subject. It has been good, too, becoming reacquainted with Mrs. Roy (Chris) Lambert, Jane Symmes, Adelaide Ponder, and others, as well as meeting Madisonians I had not known: among them octogenerian Marshall "Woody" Williams, Madison's longtime archivist/historian; talented young architect Joe Smith, and Dr. Glenn Eskew, Georgia State University history professor and resident of the Stagecoach House. This has been a personal historical celebration, indeed, for this seventy-year-old Atlanta native and ninth-generation Georgian. For me and two-hundred-year-old Madison, the timeline continues.

William R. Mitchell Jr.
February 14, 2009

Foreword

Herman Melville, who knew a thing or two about traveling, once wrote, "Life's a voyage that's homeward bound." I've been on that voyage all my life, always drawn back to Madison, always happy when I see its tree-lined streets and slow but vibrant life. And now, two hundred years after settlers founded it, this new book celebrates what makes this place special and what makes it ours.

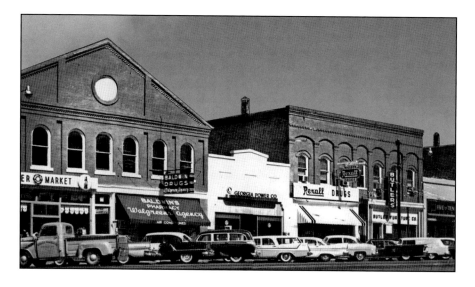

Main Street, Madison, 1956.

I won't pretend to be an impartial witness. In all, I spent twenty-six years of my life in the town, and I have visited it regularly to see my family the rest of my days. I'm so delighted that this new book has arrived to celebrate our bicentennial. And yet this isn't the first such celebration for me. I remember well Morgan County's sesquicentennial in 1957, a time when men grew beards and could be hauled off to a faux hoosegow if they were spotted on the streets without whiskers.

It's been a joyful journey for most of us.

I was three years old in the summer of 1953 when my family came to Madison in a 1939 Oldsmobile, the first car I remember. It was an era before air conditioning except for movie houses and communal freezer lockers—large centrally located freezers where people could bring food to store in those days when refrigerator freezers were about the size of a car's glove compartment. (Madison's was on Washington Street.)

I was three years old then, my brother Mark four, and we moved here because my father had taken a position as a teacher of chemistry and physics at Morgan County High School. We moved into a house near the high school on the Old Buckhead Road, a dusty dirt lane that left everything covered in an orange film in the summer.

Parade through Madison for Morgan County's 1957 sesquicentennial.

Mark and I were among the most fortunate of boys. Both our parents had served in World War II and had gone to the University of Georgia (UGA) on the GI Bill, so we grew up in a house filled with ideas, history, music, chemistry, poetry, and so much more. Six years after we arrived in Madison, our sister Laura Jane arrived, and our family was complete. We had extremely intelligent parents, and my mother was beautiful, too, and so kind and gentle to us that each day was a kind of slow glory to me. I never had better days than those gorgeous spring afternoons when I'd help Mother bring in the freshly dried clothes off the line in our backyard. No one ever had a happier childhood.

Madison was not then the town it is now. Though it was a friendly and peaceful place, and though the old homes did grace us, the downtown was like any other small southern town—pleasant but visually unexceptional. It took the hard work of many people starting in the 1960s to change Madison into the showplace it has now become.

I grew up at Morgan County High School in buildings that had been constructed decades before as part of the old Madison A&M

School. Daddy became principal halfway through the '55–'56 school year, and after that, I felt as if I owned it. To this day, there's a magic about empty buildings that I find irresistibly addictive. After school and on weekends, Mark and I had the run of the place—creating dangerous concoctions in the chemistry lab or sliding down the long metal fire-escape tubes on the sides of the main administration building.

Morgan County High School in the A&M building.

After our father left the school business, he worked in Athens, commuting for nearly twenty-five years, ending his career as an electronics design engineer at UGA. Mother opened Magic Land Kindergarten with her friend Barbara Engle, ran that for twelve years and was later a caseworker with the county's Department of Family and Children Services. When our father retired from UGA in 1986, he began to organize the county's records and wound up being the full-time county archivist for more than twenty years—never making a single cent while doing it.

All the while, my family flourished in this remarkable town. Laura Jane was the valedictorian of her class at Morgan County High School and went on to a career in nursing and teaching. Mark got his Ph.D. and became an archaeologist—one who knows as much as anyone today does about Georgia's prehistory. I became a writer and composer.

As this book demonstrates, there are thousands of such stories in Madison over its two centuries. For whites and African Americans, it has been home—that elusive place where good things—and some bad—happen to us all. Robert Frost memorably said that home is the place where, when you go to it, they have to take you in.

Madison has been taking me in all my life. The list of people in this town who have been kind to me is so long it would fill this entire book. But there is one person I cannot fail to mention with admiration and love: Adelaide Ponder. As editor of the *Madisonian*, for so many years the town's great newspaper, she was magnificent. During troubled times, her steady hand was crucial. And as my mentor and the publisher of my first journalism, fiction, and poetry, she was generous beyond my comprehension.

And yet her splendid life somehow joins with those of many such women and men who have made this town so different for so long. Black and white, rich and poor, all of us who are and who have been Madisonians have always understood this place is not like other towns of a similar size in the South.

Of course, its love of history is ever present, and that love has occasionally been at the expense of truth. Still, there has often been someone standing by to tell us what *really* happened. Myths die hard—Madison as the town too beautiful for Sherman to burn is a myth that should be relegated to the scrap heap forever here in our bicentennial—but in the end, truth will out.

And that truth is far, *far* more than enough. Alex Haley, who wrote *Roots*, said with power *and* truth that, "In every conceivable manner, the family is link to our past, bridge to our future." So true. While we look at houses and buildings and governments when we look at the history of a place, it is *family* that is central—the people who made and make this grand village what it is.

This is one book of our days. There are many others. Read it to find out who we are and the way we were. There's no place like Madison.

Philip Lee Williams is the author of fourteen published books and winner of numerous literary awards. He lives now with his family in Oconee County, Georgia.

Introduction

In 1733 when James Oglethorpe, English aristocrat and Georgia trustee, founded Savannah and the colony, the land where Madison and Morgan County came into being was still Creek Indian territory. It would remain so until the land cession and lottery of 1802–5. In this ceded piedmont plateau territory, located in the fertile area between the Oconee and Ocmulgee rivers, the state established Milledgeville in 1804 as the site for the new state capital. Georgia was expanding west, away from the Savannah River and the Atlantic Ocean, and between 1805 and 1809 ten counties were created from the ceded lands. The act establishing Morgan County came December 10, 1807. It called for a county seat with a courthouse and jail and provided for the sale of lots. The county seat, with 150 acres and a town spring, was named Madison on December 22, 1808, and incorporated December 12, 1809.

President James Madison.

This piedmont county-seat crossroads was the first new town in any state named for Virginian and Founding Father James Madison (1751–1836) after he was elected the fourth president of the United States in November 1808. (One town in New York changed its name to Madison about the same time.)

Washington, Georgia, fifty miles northeast in Wilkes County, in 1777 was the first town in America named for George Washington, a fact often repeated in these parts. But that Madison, in nearby Morgan County, was the first town in the United States to honor President James Madison at its founding has usually not been noted until now. We do so as we celebrate the two-hundredth birthday of the lovely town of Madison, looking back and tracing the events and patterns that have brought us to the year 2009.

In the 1960s, during the Civil War centennial, it became the rage to speak of Madison as "the town Sherman refused to burn." Now, in the bicentennial of Madison's founding, we offer a different thought to emphasize: Madison was the first new town to be named for President Madison, called the father of the United States Constitution and a principal author of the Bill of Rights.

Furthermore, there is a bit more justification for this celebratory fact than that old Stay and See Georgia slogan (and myth) about General Sherman's supposed refusal to burn the pretty village during his march through Georgia in the summer and fall of 1864. In fact, General Sherman himself did not actually pass through Madison that November; instead it was General H. W. Slocum. It is true that a Connecticut Yankee with Sherman's army, Sergeant Rufus Mead Jr., wrote in his diary, November 19, 1864: "Madison is the prettiest village I've seen in the state. One garden and yard I never saw excelled, even in Connecticut." That beauty is still true to this day, and indeed, the town's continuing glory is that succeeding generations have felt just as Sergeant Mead did in 1864 and have continued the process of preservation, enhancement, and conservation—loving the place—so that we may now truly celebrate two hundred years of its history through this publication.

To highlight people, events, and architectural landmarks important to Madison's evolution, we present a time line accented with interesting first-person observations and archival images. Following, in a richly illustrated portfolio, are the houses, large and small, old and new, which grace the tree-shaded streets and country roads, a rich legacy from generations of citizens who have loved this southern hometown and its surrounding countryside. Through the years street names have changed, sometimes more than once, so we are using addresses in the timeline and portfolio consistent with current usage.

But first we give an overview of Madison in the context of its development from a tiny hamlet in the piedmont frontier to a still small but energetic village today with an obvious sense of history and tradition. Beginning with Madison's place on Sherman's notorious March to the Sea, we investigate the long-held legend of this quintessential antebellum town. We then discuss the siting of the village and its town plan; we offer an analysis of the overall architecture and landmarks and emphasize the way historic preservation has fortunately played a vital role in community development, especially in the late twentieth century. We cover aspects of education, always important in Madison. The Madison-Morgan Cultural Center, for example, which started life in 1895 as a handsome brick public grade school, was transformed in the 1970s into its current place in town life; it is truly a cultural center at the heart of the village, a landmark of Madison education, and a visual monument on South Main Street to historic preservation values for all to experience.

We also include the historic churches of Madison, their place in Madison's Black and white communities, and the ongoing progress of local race relations. Here, too, is the Morgan County African-American Museum in the Moore house. The museum is next to the Round Bowl Spring Park, containing one of the local springs which had long refreshed native travelers through the area and was a significant attraction to the original white settlers.

This essay on historical development also features a section of well-known families and citizens, including noted men and women of arts and letters, commerce, and politics. Generations of Madisonians are buried in the historic Madison cemetery, including such luminaries as Senator Joshua Hill, mayor of Madison during its occupation by Union troops during the Civil War. Covered also is the *Madisonian* newspaper, which traced its origins to the 1840s, published for a time by author and journalist William Tappan Thompson, one of the best known of the town's writers, whose contributions appear on the state historical marker in front of the old *Madisonian* office on East Jefferson Street. There are county towns, such as Rutledge, Bostwick, and Buckhead, and county features, such as Hard Labor Creek State Park, an FDR New Deal creation from 1933—a long-lasting stimulus left from the Great Depression—still one of the largest of Georgia's state parks and home to a fine public golf course. Also is nearby Lake Oconee, a private development that helps to give the region the name Georgia's Lake Country.

From the earliest days, the development of Madison/Morgan County would not have been the same without its crossroads character. The stagecoach road from Charleston to New Orleans passed through the town, then narrow local and state roads, later federal highway 441/129, and still later Interstate 20, which, one is thankful, passed only close by, but not through! The Antebellum Trail, a one-hundred-mile trek designated for 441/129, begun in 1984–85, includes Madison among six other regional cities and towns, from Athens to Macon.

Fourth of July celebration at the Madison-Morgan Cultural Center.

We cannot overlook the significant place that the Georgia Railroad began to play in Madison's development in the 1840s. In fact, it was the railroad that led Union General Slocum to Madison in November 1864, as he headed for Savannah. (Although General Sherman did not come to Madison on the March to the Sea, he did pass through on the train from Augusta in February 1844 on the way to Marietta, Georgia.) Over the years, tourism, too, benefited from the highways and railroad. Actually, tourism has brought new citizens to the Madison community, as we shall show in the home profiles that follow.

We present two hundred years of Madison's evolution, from the pioneer settlement and antebellum Old South through the nineteenth and twentieth centuries onward, as we celebrate in 2009. The town has always been proud of its history and way of life, its long consciousness of living history, day by day, and of having survived the national tragedy of the Civil War to live a new day, a town surely too pretty to burn!

The Town Sherman Refused to Burn—Fact or Fiction?

In Georgia, when people of a certain age say "before the war," we mean before the War between the States, the regionally accepted way to refer to the Civil War (1861–65). Until General Sherman's strategically crucial Georgia Campaign in the summer and fall of 1864, Georgians had sacrificed sons, fathers, brothers, and husbands but had not come face to face with the Yankees in their own homes and yards. Before that, the war had been taking place elsewhere in the South, some of it even in the North, as at Gettysburg, Pennsylvania.

The extraordinary novel *Gone with the Wind* (1936) is set of course in Georgia, telling the story vividly, seemingly for all time, and it is as much about after the war as before, and not just as a glorification of the Old South and the Lost Cause. In fact, *Gone with the Wind* might as well be about Madison instead of Jonesboro, Georgia, because Federal troops marched through the town and county in mid-November 1864, as the Left Wing of the Union forces, led by General H. W. Slocum, followed the Georgia Railroad from Atlanta to Madison before turning south to Savannah. William Tecumseh Sherman, with the Right Wing, did not himself come to Madison, but to Milledgeville, the antebellum state capital, as he headed for the sea.

As Sherman was making plans to leave Atlanta on his infamous march, he sent General Grant a letter saying that he would "make Georgia howl" (October 9, 1864). He cut loose from his base of supplies and marched his entire army (some 60,000 men) toward the Georgia coast, foraging off the rich farmland of plantation Georgia. He planned to discourage Georgians from continuing to be high-spirited rebels, bring the war to their yards, gardens, and smokehouses, and cut a debilitating swath across the Confederate landscape.

It has always been said that Sherman's army was careless with fire, especially in Atlanta. Some years after the war, *Atlanta Consti-*

General William T. Sherman at Atlanta.

tution editor Henry W. Grady told Sherman to his face in a speech that the general had been "a might careless with matches." Grady intended it as a joke, and it was so received, with howls of laughter.

It can be said on General Sherman's behalf, however, that he issued a field order, number 120, November 9, 1864, that contained a degree of consideration for the citizens as his troops marched, but even Sherman admitted that his order was honored more in the breach than the observance. In part Section IV of the field order read, "the army will forage liberally on the country during the march. Soldiers must not enter the dwellings of the inhabitants or commit any trespass, but may be permitted to gather . . . provisions and forage at any distance from the road traveled." Section V had this: "To any corps commander alone is entrusted the power to destroy mills, houses, cotton gins, etc." It read further, "when the army is unmolested no destruction of such property should be permitted." Section VI allowed the "free appropriation, without limit, of horses, mules, wagons, etc." It read, "In all foraging of whatever kind, the parties engaged will refrain from abusive or threatening language . . . and they will endeavor to leave each family a reasonable portion for their maintenance." After the war in a graduation address at a military academy, Sherman said, realistically, "War is at best barbarism. . . . War is hell" (June 19, 1879).

One question before us here is how much "hell" did Madison and Morgan County experience in mid-November 1864, when the war came to this place in the Georgia piedmont that had been in existence for only two generations—fifty-five years? Was it any less devastating here than in nearby Covington, Eatonton, Social Circle, Milledgeville (the state capital), or Savannah, the ultimate target in Georgia as the army headed for South Carolina? Did Madisonian Joshua Hill, a staunch Southern Unionist and Madison's mayor at the time, in fact protect Madison any more than Field Order 120? Madison has long been famous as the town Sherman refused to burn because it was too pretty. Today it is still considered essentially a preserved antebellum town along the Antebellum Trail from Athens to Macon.

The transition from the Old South, before the war, to the New South afterward has been an ongoing, oft-told story, much of it romance, myths and legends, novelistic interpretations, tall tales, folk-

ity. In the article, Elissa R. Henken, a University of Georgia folklorist, is quoted as saying, "Each town [on the march] treats itself as the one and only that was saved." This 2004 article closed with "Lore of Madison." McWhirter wrote: "Madison, a hamlet of well-preserved antebellum homes in east Georgia, has at least six legends about how the town was spared Sherman's wrath. The most popular is that Sherman found the town 'too beautiful to burn.'"

Madison itself published a leaflet in the early 1970s, about the time

Federal forces moving out of Atlanta on the March to the Sea.

lore, well-meaning chamber of commerce public relations, and tourism publicity. A fairly well-balanced account is an *Atlanta Journal-Constitution* article by Cameron McWhirter on Sunday, May 9, 2004, headlined "Sherman Still Burns Atlanta." The writer says that it was "one of the most famous military campaigns in American history . . . still taught at West Point as an example of how to break an enemy's will to fight." One expert academician interviewed said that Sherman's soldiers "could have done a lot worse that they did." Sherman himself is quoted from his memoirs as saying that in Atlanta he was "after military targets, 'fair game,' near the depot, but the fires did not reach parts of Atlanta where the courthouse was or the great mass of dwelling houses." And there was some truth in that—the Fulton County courthouse survived, as did a number of mansions and churches in that vicin-

Interstate 20 was being completed through Georgia. The Madisonian Printing Company, Inc., produced the fold-over leaflet for the Stay and See Georgia program of the state's Department of Industry, Trade, and Tourism; the leaflet's headline was "The Town Sherman Refused to Burn . . . Madison, Georgia." The story is repeated that "Sherman spared Madison on a personal plea from [former Congressman] Joshua Hill, a southerner who had not voted for secession." It continues: "Madison is famous for its beautiful tree-lined streets graced by some of the handsomest old homes in the South. The town has also retained much of the culture and charm that compelled White's *Statistics of Georgia* in antebellum days to call it 'the most cultured and aristocratic on the stage coach route from Charleston to New Orleans.'" (The stagecoach might be long gone but the official story has pretty

One day, two citizens, Messrs. Hill and Nelson, came into our lines at Decatur and were sent to my headquarters. They represented themselves as former members of Congress and particular friends of my brother John Sherman; that Mr. Hill had a son killed in the rebel army as it fell back before us somewhere near Cassville, and they wanted to obtain the body, having learned from a comrade where it was buried. I gave them permission to go by rail to the rear, with a note to the commanding officer, General John E. Smith at Cartersville requiring him to furnish them with an escort and an ambulance for the purpose. I invited them to take dinner with our mess, and we naturally ran into a general conversation about politics and the devastation and ruin caused by the war. They had seen a part of the country over which the army had passed and could easily apply its measure of desolation to the remainder of the State, if necessity should compel us to go ahead.

Mr. Hill resided at Madison, on the main road to Augusta, and seemed to realize fully the danger; said that further resistance on the part of the South was madness, that he hoped Governor Brown of Georgia would so proclaim it and withdraw his people from the rebellion. . . . I told him, if he saw Governor Brown, to describe to him fully what he had seen and to say that if he remained inert, I would be compelled to go ahead, devastating the State in its whole length and breadth; that there was no adequate force to stop us, etc.; but if he would issue his proclamation withdrawing his State troops from the armies of the Confederacy, I would spare the State, and in our passage across it confine the troops to the main roads and would, moreover, pay for all corn and food we needed.

William Tecumseh Sherman
From his memoir, first published in 1875. (Lane, 1974)

much remained: Stay and See Madison.) "It is accessible from most major highways. Located on highways 441 and Interstate 20, it is one of the most delightful tourist routes going north and south east or west." This now somewhat rare leaflet lists about fifty mostly antebellum sites, keyed to a small map, and all of the sites are in this book.

Probably the final word on the subject came in the summer 2002 issue of the *Georgia Historical Quarterly*; Brian Melton, a graduate student, published a twenty-nine-page article, "The Town that Sherman Wouldn't Burn: Sherman's March and Madison, Georgia, in History, Memory and Legend." (Melton partly based his scholarly article on interviews with Marshall W. Williams, Madison historian and archivist, and on an article by Williams from 1997 in *Morgan County Heritage, 1807–1997*.) Melton points out that "Madison lay on the Georgia Railroad and was the last major stop before the line reached Augusta. This Madison location set it on the path of Sherman's army when it left Atlanta on November 15, 1864." Melton acknowledges that Hill did meet with Sherman in Decatur, near Atlanta, but doubts he did so to specifically intercede on Madison's behalf, or that guards were posted at each house in Madison because of Hill's pleas. (In fact, guards were often provided when men were available for such duty to keep at bay the so-called "bummers," stragglers, who did much of pillaging along the way. (The diary of Mrs. Dolly Sumner Lunt Burge, a sometime Madisonian who was a resident of the plantation country

Union troops destroying factories and rail lines in Atlanta.

Federal soldiers burning the Madison rail station.

of nearby Covington in Newton County, describes such guards whom she got to know and got along with, to a degree. She was herself originally from Maine, but considered herself a Confederate.) Melton further points out that Madison wasn't really "occupied" since the army took only two days, November 18–19, to pass through the town.

Sherman's memoirs refer to a meeting in Decatur with the Southern Unionist Hill, but the primary reason was for Hill to seek permission to retrieve his son's body from a battlefield north of Atlanta. Sherman noted that Hill felt "further resistance on the part of the South was madness," and that he hoped to convince Governor Joe Brown to withdraw Georgia from the Confederacy. Sherman wrote he assured Hill that if Brown withdrew the Georgia troops, he would "spare the State," keep his army on the main roads, and "pay for all the corn and food we needed." Brown was apparently tempted, but the opportunity passed.

Melton's article details what military targets were destroyed in Madison on November 19, 1864, including Georgia Railroad tracks and the depot in town. There was some ransacking in the downtown area. Some cotton bales were torched. The Madison Steam Mill, which was involved in the war effort in various ways, was partially damaged by fire, but it is believed the townspeople put out the fire. It was known in town that Hill had met with Sherman, and that Hill, a man of reputation and influence, had opposed the war and feared the certainty of its consequences. Melton writes, "From the perspective of the average resident of Madison it could easily appear that [Mayor] Hill himself saved the town."

Melton quotes diary accounts of a number of Union soldiers who were "very impressed with the town." One soldier, a New Yorker named McMahon, said "Madison was the best looking place I have ever seen in the whole South." (There are several of these references, including that of the Connecticut Yankee Sergeant Mead quoted earlier.) Melton closed his article on the question of Madison's history and legend—exactly why had the town been spared? "In Madison, as well as other Georgia towns, the debate will rage for years to come."

As time passes the question may fade, but the thing that will not be debated is that the pretty two-hundred-year-old town will continue to be preserved and thrive. The epitaph on Joshua Hill's burial slab in the old Madison Cemetery reads, "A Staunch Southern Friend of the Union." Hill can rest in peace that his Unionist views at least did nothing to harm the place and may have indeed helped—he was the mayor, after all.

Madison was not an obscure backwater during the War between the States, when it already had a reputation for culture and beauty. It still has that reputation, and it remains a beautiful place today. Ironically, its reputation may have been spread by having been on Sherman's March.

Town Plan, Early Architecture, and Landmarks

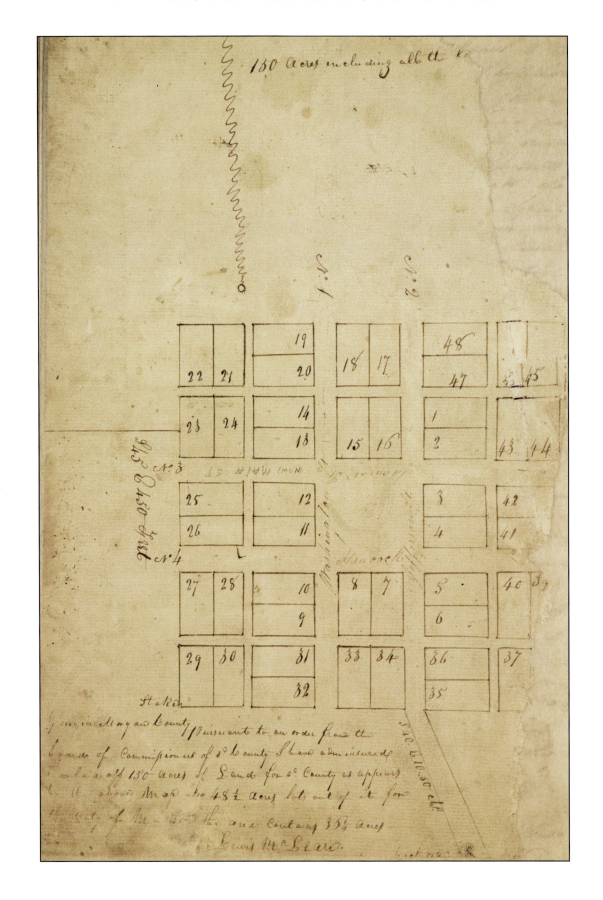

The 1845 Morgan County Courthouse, about 1907.

The current Morgan County Courthouse, about 1907.

Looking back two hundred years, it is apparent that the pattern of the original town plan, with four principal streets, Monroe (now Main), Jefferson, Washington, and Hancock, a grid around a central square, has changed little since it was designed in January 1809. Forty-eight town lots around the town square were sold to provide funds to construct a courthouse and jail. The lots surrounded the square two blocks deep and were purchased by thirty-three different men about whom, with few exceptions, little is known. One of these, Reuben Rogers, built on lot number six in 1809–10 the two-story frame house that today is the Rogers House Museum. Now restored on its original site, it stands adjacent to the present courthouse, on Jefferson Street, and is an example of the Piedmont Plain style of architecture, which was the basic root stock for antebellum residential design in this area, preceding the Greek Revival and the less-popular Italianate style.

The Madison plan, with the central courthouse layout, was similar to that in Washington, in Wilkes County, and was also used in nearby Eatonton and Monticello. The justices of the inferior court had purchased 150 acres of land for the town, which is at the center of the county, and laid out those original forty-eight lots. (Reuben Rogers, for example, paid $111 for lot number six, selling it in 1811 for $1,000, after he built his house.)

The courthouse stood in the center of the public square, now Post Office Square. This first building, begun around 1810, burned in 1844, although the county records were saved. In 1845 the second courthouse was constructed (described in 1849 by George White as a "spacious brick building") and served until larger space was needed. The present courthouse was built in 1905–7 northeast of the square on Jefferson Street. The old courthouse in the square was used for various purposes until it, too, was destroyed by fire in 1916. The present domed, Neoclassical Revival, red-brick "pile" was designed by a well-known early-twentieth-century Atlanta firm of architects, J. W. Golucke and Company. It cost about $40,000, a handsome sum in the first decade of the last century. (James Wingfield Golucke, 1857–1907, a native of Crawfordville, Georgia, was a self-taught architect who specialized in county courthouse design; many of his twenty documented Georgia courthouses are still in use, including the ones his firm did in 1905 for Putnam, Worth, and Morgan counties.)

Introducing The *Madison Historic Preservation Manual*, published in 1990, Dr. William Chapman, then of the University of Georgia, wrote: "Madison is . . . widely recognized for its outstanding historic architecture and its overall aesthetic and environmental qualities. First listed [as a district] on the National Register of Historic Places in 1974 (with further expansion of the listing in 1990), Madison stands as an unequaled reminder of the character and quality of life in the Georgia piedmont over the last two hundred years." The Madison district nomination says the place is "an intact, well-preserved town characterized by a commercial district in the center, with adjacent residential neighborhoods to the north and south, which are dominated by antebellum architecture." It mentions a Victorian section to the east and a city cemetery in the southwest section. "West of the square are commercial and industrial buildings, particularly along the railroad."

What could be more basic than Madison's well-preserved original layout as we proceed through the development of the town from its founding? This book traces how the place got to be the way it is today in its bicentennial—its architectural development, spreading out from the square to the points of the compass along roads that came into being as the town became a crossroads in the center of Morgan County, which it remains to this day. The old 1809 drawing of the central square and town plan shows a spring, now called Round Bowl Spring Park, as a naive squiggle, just a few blocks west of the square, and this little oasis was clearly one of the reasons for the town's location.

After its founding, the town quickly grew and thrived, but not as much as it would later when the cotton economy boomed in the 1830s and '40s. As the county seat became the locus of trade and transportation, churches, social life, and education, the city limits were extended in 1822 to include, the act read, "all land within one-half mile of the public square." Early on, Madison became a stop for travellers en route along the main wagon road from Philadelphia via Charleston to New Orleans. Part of the route, the Seven Islands, or Alabama Road, was an important part of the Upper Creek Trading Path, long before the Creek Indian cession of 1802–5. Traffic on this road crossed the Oconee River on Park's Bridge, which was opened in 1807 near Buckhead.

Round Bowl Spring Park.

From the 1820s to the 1850s, large residential lots were developed with fine houses, many of them in the Greek Revival style. This new growth away from the original town layout, where the Plantation Plain style Rodgers house was built, occurred along the north/south Main Street corridor and adja-

Reuben Rogers house (1809–10).

Boxwood in 1981 with a view of the 1844 Gothic Revival Church of the Advent across Academy Street.

cent streets, and most of it survives to this day. Three-story Boxwood, built in 1851 on Old Post Road, is a great mansion that combines the Greek Revival and Italianate styles in an extraordinary boxwood parterre garden setting, front and rear. (A cultural geographer, Wilbur Zelinsky, showed in 1954 in a scholarly article about antebellum Georgia towns that it was in the towns, rather than the countryside, that the great mansions of the cotton planters, such as Boxwood, were built.) Madison has been largely known as a beautiful, surviving, antebellum, *White Columns in Georgia* town. (In the 1952 book by that title, there is a long chapter on Madison; see bibliography.) Little more evidence would be needed for that reputation than Boxwood, which spans the depth of a city block, its formal Greek Revival façade looking out on Old Post Road and its long, shaded Italianate porch facing Academy Street.

But in addition to such mansions that are evidence for Madison's reputation as a white-columned, antebellum town, there are good ex-

Two large gardens, where peacocks fan their tails in the green aisles of boxwood, give their names to Miss Newton's place. . . . In Madison you can almost feel that you are back in the Old South.

A Guide to Early American Homes: South (1957).

amples of the pre–Greek Revival, two-story Plantation Plain style. These are along the old streets, Academy, Old Post Road, and Dixie Avenue, west of South Main Street: the Stagecoach House on Old Post Road combines frame Plain style features from 1810–20 with c.1840 Greek Revival aspects such as a columned porch, added when it is thought to have served as a stagecoach stop. On Academy Street is the Edmund Walker town house, 1838, which is pure Plain style. Thurleston, on Dixie Avenue, is a high-style mansion that combines Greek Revival with suggestions of the Gothic Revival, yet its earliest

TOWN PLAN, EARLY ARCHITECTURE, AND LANDMARKS

Above: View of downtown Madison from the courthouse bell tower in 2009 with the town square and post office in center.
Below: Plain style Rogers-Shields house on North Main Street.

section dates from 1818 in the Plantation Plain style; an old brick chimney and the roofline of the 1818 house can be seen clearly at the rear on the north side of the structure. There are also early houses on North Main Street; the frame Rogers-Shields house, for example. Although a large house, it is basically Plain style; in fact, it started life as a double log cabin with an open dog-trot breezeway in the center of the plan.

In 1982, this writer observed in *Landmark Homes of Georgia, 1733–1983*: "Madison was one of the most prosperous of the Piedmont towns; by the 1850s it was the center of a civilized way of life similar in many ways to that in the rest of the Union, with the exception that the labor force was in servitude." This was the village that Sherman's Federal Army encountered in November 1864 on its March to the Sea, as it followed the Georgia Railroad from Atlanta east. A downtown fire in 1869, only four years after the war, was far worse than anything the Federal army did in the fall of 1864. The 1869 fire destroyed forty-two structures and all but one business, but the postbellum recovery from this urban renewal was complete by the 1870s and '80s. A century later the city of Madison began several new initiatives to encourage investment in a deteriorating downtown. Through the creation of a local Main Street Program and a Downtown Development Authority, Madison has been able to revitalize its commercial core, and today it has a bustling town center with a variety of restaurants, galleries, shops, and offices in the handsome brick buildings from the nineteenth century.

Madison, like many southern towns, obviously values its architectural and social history, but it is truly remarkable that a town with a population of only 4,400 would have in its midst six museums—the Rogers House, Rose Cottage, Heritage Hall, the Richter Cottage, the Morgan County African-American Museum, and the Madison-Morgan County Cultural Center. In fact, the entire Madison Historic District is essentially a museum of nineteenth-century architecture, but it is a *living* history museum, as this is a thriving community, though assuredly a scenic destination as well, for tourists and prospective townsfolk.

Education and the Madison-Morgan Cultural Center

Located in the heart of the National Register historic preservation district, the Madison-Morgan Cultural Center, 434 South Main Street, is a living symbol of the continuing educational values in history of Madison. The center, which was opened in July 1976 by the Morgan County Foundation, Inc., represents and gives continuity in community education from generation to generation, as it helps educate new generations. The imposing red-brick building was a graded grammar school opened in 1895, not long after Georgia instituted a system of tuition-free public schools. When it was new, it was in the latest High Victorian style, the Romanesque Revival, with characteristic rounded arches, a central bell tower, and touches of classical details. The rear auditorium wing, with a conical roof, is especially handsome.

Within walking distance south of the courthouse square, and with plentiful parking and spacious grounds, the cultural center is a community destination for educational exhibits and events, including an annual July Fourth celebration (held on July 3!). After the building ceased to be a school, the town library was located there, beginning in 1947, where it stayed until 1975 when the cultural center project was underway and the library was moved to the present East Avenue location. It had been named the Uncle Remus Regional Library in 1958. (Uncle Remus was the fictional character created by Atlanta writer Joel Chandler Harris [1845–1908], who was born in nearby Eatonton.)

Above: Fourth of July Celebration at the Madison-Morgan Cultural Center. Below: Madison High School (1921).

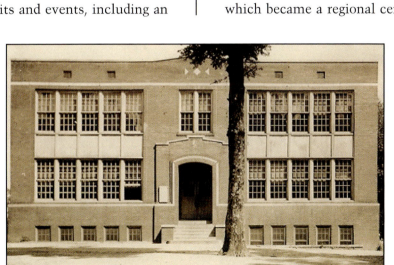

The Reverend George White is often quoted about antebellum Madison; in his *Statistics of the State of Georgia* (1849), he said: "Madison has long been celebrated for excellent schools. There are as many well-educated gentlemen and ladies in Madison as in any portion of the state." Also symbolic of Madison's place in education from its earliest days is a monument on courthouse square, the Braswell obelisk honoring Benjamin Braswell, who in 1817 established an educational trust to aid the children of destitute widows. The trust in 2009 is still being used as intended.

In 1810 a Madison male academy was begun in a two-story brick building, which burned in 1824. In 1820 it had seventy-four students studying the three r's, plus grammar, geography, and classical languages. A female academy was also founded early on, as well as several other private schools for girls. This established a pattern in Madison, which became a regional center for female education, including two colleges by the 1850s. In 1871 a board of education was elected. Most of the county schools for white students grew out of the existing private schools. In 1909 the county voted countywide taxation to support public schools, and in 1948 the white high schools in the county were merged into one. Morgan County schools were integrated in the 1970–71 school year, and all the county schools are now consolidated in Madi-

The Eighth District Agricultural and Mechanical School, established in 1908.

son, divided into high, middle, elementary, and primary.

Although it ceased to exist in 1933, the Eighth District Agricultural and Mechanical School was established in Madison in 1908, part of a system of twelve such institutions established by the Georgia legislature. Its campus was located on the east side of Madison on three hundred acres that are the present site of the high school. There was a farm with a dairy barn, farm animals, and fruit trees. There were classes, dormitories, literary societies, and athletic teams. Mr. Toombs Kay taught horticulture; one of his children is author Terry Kay. Madison A&M is mentioned prominently in his *To Dance with the White Dog* (1990), his best-selling novel that was made into a Hallmark Hall of Fame television movie in 1997.

Two colleges for female students, sometimes described as finishing schools, gave a refined tone to antebellum Madison. The first of these was founded as the Madison Collegiate Institute in 1850 by Baptists led by John Byne Walker, a wealthy planter. When it was incorporated as the Georgia Female College in 1851, the prospectus read, "The town of Madison is synonymous with wealth, refinement, and morality." Students would board at the homes of trustees and other leading Madisonians. The Greek Revival style Baldwin-Williford house, 427 South Main Street, remains from the college, which permanently closed around 1882, after struggling during the post-war years.

A four-year Methodist institution was incorporated January 26, 1850, the Madison Female College. In 1854 the Reverend George White in his *Historical Collections of Georgia* described it as having "a course of study that embraces every useful and ornamental branch." The campus was on Academy Street, near the present site of the Episcopal Church of the Advent parish house. In the early 1860s, as the war raged, the Madison Female College closed, and the main college building (which burned in 1864) was used as a Confederate hospital. Charles

An engraving of the Georgia Female College. The Baldwin-Williford-Ruffin house, now a private residence, is on the left.

Edgeworth Jones of Augusta wrote in his *Education in Georgia* (1889): "A curriculum of high order was in force. . . . The faculty, generally, comprised some eight or ten members, and an annual attendance averaged about 150 pupils."

For the year 1851–52, Miss Rebecca Latimer, a senior from Decatur, Georgia, boarded with the college president on the Old Post Road. At her commencement in 1852, she met Dr. William A. Felton, whom she married in 1853. Later, Dr. Felton became a U.S. congressman. At age 87, Rebecca Latimer Felton was appointed a U.S. senator from Georgia, the first woman to occupy a seat in the United States Senate. She is considered the most notable graduate of the Madison Female College.

Before the war, several boys' academies existed in Madison, including one in which Alexander Hamilton Stephens, later vice president of the Confederacy, taught in the year 1832 after he graduated from the University of Georgia. Stephens boarded for a time in the Cooke house, 287 Academy Street, an early Greek Revival style house, the earliest part of which dates back to 1819. A 2008 novel, *The Gates of Trevalyan*, by Jacquelyn Cook, begins with Alex Stephens teaching in Madison and his legendary romance with a young lady named Sarah Allen. Stephens is just one of many well-known people who passed through Madison because it was an educational center, and today it is once again receiving state and national recognition for its public schools.

Restored school room from the Madison Graded School on display at the Madison-Morgan Cultural Center. A room like this would have been where young Norvell (Oliver) Hardy attended first grade in the 1898–99 school year.

Madison's Historic Churches and Congregations

The antebellum churches of Madison. Above on Main Street: The 1858 Madison Baptist Church in the foreground and the 1842 Madison Presbyterian Church in the distance. Opposite page: The Madison Presbyterian Church and a Gothic Revival church, built by the Methodist congregation in 1844 and renovated in 1961 to house the Episcopal Church of the Advent.

Antebellum Madison had four major protestant denominations, Baptist, Episcopal, Methodist, and Presbyterian, and a small Roman Catholic congregation. At 382 South Main is the Presbyterian Church, built in 1842 in a simplified Greek Revival style. The builder was Daniel Killian, a master mason. It had a slave gallery that was entered from the outside. In 1908 Elizabeth Speed donated seven Tiffany stained-glass windows for the church. Distinguished Presbyterians Alexander H. Stephens and Ellen Axson worshiped here. Ellen Axson's father was the church's minister just after the Civil War, and later she became President Woodrow Wilson's first wife.

The Baptists built a frame church in 1834 on Academy Street, then in 1858 they constructed a classical meeting house at 328 South Main. There were 267 members of the church: 147 white and 120 black, who in those antebellum days sat in a slave gallery, still preserved as a historic record. The bricks for the church are said to have been made on the Morgan County plantation of leading Baptist layman John Byne Walker, who helped found the Georgia Female College nearby. The church has undergone significant architectural changes over the years. The steeple was remodeled in 1862 and several times thereafter, and in 1917 a classical portico and outside steps were added to the front.

The first Methodist meeting house in Madison was frame and built about 1825 on Academy Street. In 1844 the Methodists replaced it with a brick Gothic Revival church. The new sanctuary was dedi-

cated by Bishop James Osgood Andrew, who had been assigned the Madison/Athens circuit in 1829 and was a founding trustee of Emory College at nearby Oxford in 1836. Andrew had been made bishop in 1832, and by 1844 his ownership of slaves became the catalyst that brought to a head the simmering disagreement between slaveholding and anti-slave factions in the Methodist faith, resulting in the formation of the Methodist Episcopal Church, South, which remained a separate entity until 1939.

The present United Methodist Church, a large neoclassical style cream brick structure, was built in 1914 across South Main Street

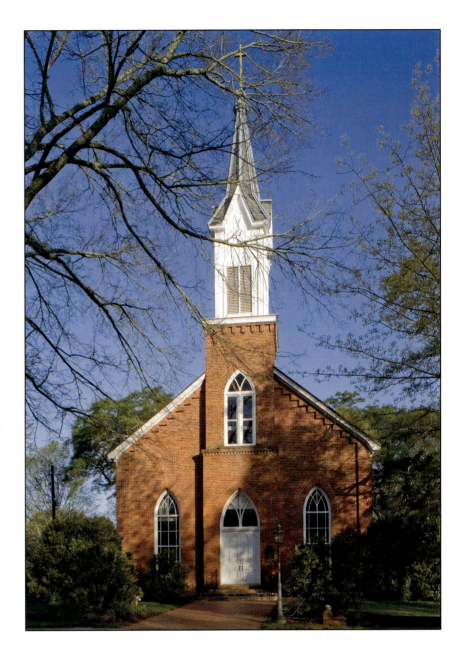

from the Baptist and Presbyterian churches. The 1844 church building is today the Episcopal Church of the Advent.

About the time the other denominations were building along South Main Street, the Episcopalians began worshipping in a small frame building, no longer standing, and in the early 1850s built a Greek Revival church just inside the grounds of the old cemetery off Academy Street. That church was deconsecrated in 1937 and later razed. The congregation remained intact, however, and purchased the former Methodist church on Academy Street, which was renovated in 1961 and consecrated in 1963.

Within the Madison Historic District is an early frame church with an interesting history. Built on Academy Street in 1834 by the Baptists, the building began being used in 1866 by the first independent Black congregation in Madison, Calvary Baptist. The next year it was rented by the Freedmen's Bureau for use as the first school for Black children in Madison. The bureau then purchased the building, disassembled it, and moved it to the west end of Hill Street where it was reconfigured and used as a church and school. In 1883, when the Calvary Baptist congregation completed a new brick church on Academy Street, another congregation, led by Reverend Allen Clark, was established and has met there ever since. It is known as Clark's Chapel Baptist Church.

The Saint Paul African Methodist Episcopal Church, 811 Fifth Street in the Black residential area called Canaan, dates from 1881. A red-brick, white-steepled landmark on the Canaan landscape, it faces the railroad tracks and is an active congregation in its neighborhood.

At 156 Academy Street, adjacent to the Calvary Baptist Church, is the Morgan County African-American Museum, which opened in 1993. The frame Victorian farmhouse, built around 1900 by former slave John Wesley Moore, was donated by Reverend Alfred Murray and moved to this site where it was restored with funding by the DuPree Foundation. Madisonian and former State Archivist Carroll Hart, speaking at an opening ceremony in the church next door,

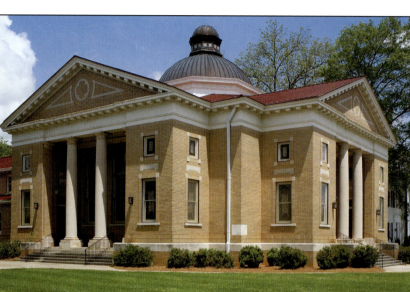

Above: United Methodist Church, built in 1914 on South Main Street. Below: Clark's Chapel Baptist Church on Hill Street.

called the museum a "miracle on Academy Street." She said it was the accomplishment of a group of community leaders, both Black and white. Notable African Americans in the leadership were Madison-born artist Benny Andrews and community leaders Fred Perriman and Martin Luther Bass. Georgia Congressman and Civil Rights leader John Lewis wrote Fred Perriman at the time of the dedication, "The opening of the African-American Museum in Madison, Georgia, is another positive step in the direction of commemorating African-American History."

Overall, race relations in Madison and Morgan County in the two-hundred-year history of this middle Georgia community have been peaceful. Although, as in much of the nation, true integration has not been attained, Black heritage and culture have added rich traditions to the place. (See Rose Cottage, built in 1891, the museum home of an ex-slave.) In 1865 the Freedmen's Bureau was established to assist former slaves with education and other necessary aspects of a better life. In 1867, when the head of the bureau rented the former Baptist church on Academy Street for a school, Black education in Madison had formally commenced.

Those were the hard, post–Civil War–Reconstruction days, but much was accomplished, and, proudly, that tradition persists, as Madison continues to evolve toward an integrated community of Black and white souls, churchgoers, keeping the faith.

*Above: Calvary Baptist Church, built in 1883 on Academy Street.
Below right: The Morgan County African-American Museum on Academy.
Below left: Saint Paul African Methodist Episcopal Church, 1881, on Fifth Street.*

Madisonians: People, Society, and Newsmakers

The Georgia General Assembly published a book about the state written in 1895 by R. T. Nesbitt, commissioner of agriculture. Nesbitt wrote in *Georgia, Her Resources and Possibilities,* "Madison, the county seat of Morgan County, is one of the most beautiful towns in Georgia.... The citizens are intelligent and known far and wide for refinement and taste in all that makes for the good of society. It is noted for the taste of ladies in the cultivation of flowers in their beautiful flower yards." (The Madison Garden Club was founded in 1895. It was the first garden club in Madison and the second oldest in Georgia [after the Ladies Garden Club in Athens]. In 1932, after some years of inactivity, it was rejuvenated and renamed La Flora.) This section is about Madisonian citizens, a selection of people and the society they helped form, and newsmakers and newspapers, most notably the *Madisonian*, which came into being in 1879. The *Madisonian* succeeded the *Southern Miscellany*, started in 1842 by publisher C. R. Hanleiter and edited from 1842 to 1845 by famous southern wordsmith William Tappan Thompson (1812–82). Thompson's classic book *Major Jones's Courtship* (1843), humorous stories about Madison and Morgan County, originally appeared as a series of letters in the *Southern Miscellany*.

In that newspaper, Thompson wrote in 1842: "We folks in Madison just love to have visitors. The newspaper likes to get your name and record it so other folks can read about you." By the 1840s, when Atlanta, for example, was founded as a railroad terminus, Madison already had a Georgia Railroad depot and was a destination, even then, for genteel tourists. On September 15, 1845, Madison and Atlanta were joined by rail, when the first train ran from Madison to newly named Atlanta, at that time hardly the urbane place it is today.

Then, however, Madison was a sophisticated town, and Thompson was just one of many talented people that were associated with the place, among them Alexander Hamilton Stephens, later the Confederate vice president, and Rebecca Latimer Felton, who as the widow Felton became the first ever female U.S. senator. Those were the sort of people and news that Madison was attracting in the nineteenth century and that it has continued to do into the next centuries.

In 1849 Reverend George White wrote in his *Statistics of the State of Georgia*: "In point of intelligence and hospitality, Madison acknowledges no superior. Many of the citizens are wealthy, and live in much style." This was confirmed by Madison teacher and Methodist Dolly Sumner Lunt Burge, originally from Maine, who wrote in her now well-known diary (1848–79) on March 30, 1849: "Dressed and went to dinner at twelve o'clock, fashionable dinner, very much so, large company present. Mr. Kolb returned from plantation to which he went yesterday." Dolly Lunt boarded with prominent citizens Wilds and Nancy Kolb, who completed Boxwood in 1851, across from the Madison Female College, a Methodist school on Academy Street where it is believed Dolly taught at that time. She later moved to nearby Newton County in the vicinity of Covington and married Thomas Burge, a plantation owner, who left her the plantation when he died in 1859. She wrote about Sher-

> *Who would live down among the sand flies and mosquitoes when the railroad brings oysters to your door in their season?*
> Southern Miscellany
> November 27, 1842

> *I seed your piece to correspondents, whar you said you hoped Majer Jones would rite for your columns, and I wanted to tell you that you mought spect to hear from me every now and then, if you likes my ritins. I felt a little sort o' scared at fust; but all my quaintances as had read my letter to you, advise me to go a-head and be a literary caracter, and as you want me to rite for the "Miscellany," I'm termined to do what I kin to rais the leterature of Pineville.*
> Major Joseph Jones
> Major Jones's Courtship (1843)

man's March to the Sea as his troops passed through her Newton County property in November 1864. She said it made her "a much stronger Rebel!"

As the mayor when Sherman's army came through town, Joshua Hill certainly played a role in the March to the Sea episode of Madison's history. Here was indeed "one of Georgia's most distinguished sons," as historian/archivist Lucian Lamar Knight described him in *Landmarks, Memorials and Legends* (1913): "United States Senator [elected 1868] Joshua Hill was long a resident of Madison; and here he lies buried. On the eve of the Civil War, Mr. Hill was a member of Congress [elected 1856]. He was not only a strong Union man, but an anti-secessionist." Knight added, "On constitutional grounds, he resigned from Congress in 1861 when Georgia seceded." As a United States senator during Reconstruction, he helped the state in many ways. Knight wrote that this Madison contrarian "in religious matters . . . was strongly inclined toward agnosticism." Knight says that Hill's "income from the practice of law was immense, and . . . he accumulated a fortune."

The Joshua Hill grave in the Madison Old Cemetery.

Hill's Madison home on Old Post Road, a white-columned mansion that combines antebellum Greek Revival and postbellum Neoclassical Revival, has a state historical marker that mentions his unsuccessful bid in 1863 for governor of Georgia. The text sums up the life of this Madisonian man of principle, "He never lost the respect and admiration of the people of Georgia." (Timothy Furlow, another Madisonian, who founded Americus, Georgia, also ran in that election; an ardent Confederate, Furlow also ran against Joseph E. Brown, who won.) Joshua Hill was definitely a Madison newsmaker, and his life and career tell us much about the historic good character of the citizens of Madison/Morgan County, Georgia.

A handsome two-story frame house at 433 South Main Street, begun in 1850 and now having a long, one-story neoclassical front porch, was the home of two of Madison's most interesting citizens, Mr. and Mrs. Paul M. Atkinson. Mrs. Atkinson had been Miss Lulu Hurst, once known as the "Georgia wonder girl," whose feats of mind-over-matter strength were legendary. Born in 1869 near Cedartown in northwest Georgia, she became, as a teenage girl, an internationally famous stage entertainer. Always chaperoned by her father, she did mysterious stunts of lifting, and one of the towns to which she brought her act was Madison. It was a local boy with a dramatic flair, Paul Atkinson (1858–1931), who introduced the slender titan to the audience. Sufficiently impressed, her father hired Atkinson to become his daughter's business manager.

They moved on to Atlanta and New York City, and she "packed them in" everywhere they went. The *New York Times* reported June 5, 1884, "The entire performance was a wonderful exhibition of an unaccountable power, and the immense audience was delighted and amazed." But she grew tired of being a traveling female curiosity and fell in love with Paul Atkinson. They married and moved to Madison to 433 South Main Street. Both became leading citizens, but Lulu, the mother of two sons and a matron, seldom spoke again of her stage career.

Paul and his brothers operated the family's Madison Variety Works. But Paul's flair for show business didn't die as easily as Lulu's. In 1890 he bought the Battle of Atlanta Cyclorama painting for $2,500 and after exhibiting it for several years sold it in 1892. In time the city of Atlanta acquired this cyclorama painting and erected for it a building where the marvelous canvas is still exhibited in Atlanta's Grant Park.

Louise Marion McHenry Hicky (1891–1984), Mrs. Daniel Hicky,

was born in the Stokes-McHenry house, 458 Old Post Road, where generations of her family had lived since 1822. In 1948 she started for the *Madisonian* newspaper a column called "By the Way," which she continued for fifteen years. In 1971 her book *Rambles Through Morgan County* was published for the Morgan County Historical Society.

A new edition, published in 1989, has been the source of much of the information about the Madisonians Paul and Lulu Atkinson. At the time Mrs. Hicky published her *Rambles*, she lived in her family home where she was born. In 1978 Louise Hicky was featured in the beginning of a two-issue *New Yorker* magazine article by E. J. Kahn Jr., "Georgia, From Rabun Gap to Tybee Light," also published as a book. Kahn quoted Louise Hicky: "We've never had the pleasure of selecting our own furniture." When she referred to "the War," Kahn wrote, "she means the War between the States." The house is known for the family collection of old manuscripts and first editions. One manuscript that is especially treasured is Mrs. Hicky's grandmother's diary of a European grand tour in 1859, sailing from New York in May and returning to Madison in the fall of that year.

The *Madisonian*, for which Mrs. Hicky wrote her column, had been as much a part of Madison since 1879 as Round Bowl Spring, Bonar Hall, or Boxwood, or, for that matter, the Stokes-McHenry house with its subtle combination of graceful and authentic Greek Revival and Gothic Revival styles. When Louise Hicky began her "By the Way" pieces in 1948, the *Madisonian* publisher was Edward Taylor Newton II (1904–83), a lawyer and scion of the Madison Newtons. When he died in 1983, the *Madisonian* obituary of January 7 read, "He was a former owner and editor of *The Madisonian*, published by his family for more than fifty years." Newton was the coauthor of the book *The Fundamentals of Patent Technology*. The obituary ended with a mention of his sister, Miss Therese Newton of Madison.

Ellen Therese Newton (1905–96) was an educator who resided at Bonar Hall on Dixie Avenue. The second husband of her mother, Josie Varner Newton, was William T. Bacon. He edited and owned the *Madisonian* from 1894 until 1944, and Bonar Hall, one of the major landmark houses and gardens of Madison, was their home. Therese Newton inherited Bonar Hall, and it is still in the Newton family.

The *Madisonian* passed through several ownerships until the husband and wife team W. Graham and Adelaide Ponder, both Madison natives, took over in 1957 as publisher and editor. They operated the enterprise from 131 East Jefferson Street, a two-story brick office building that they renovated. The Ponders and the historic newspaper became a symbiotic part of Madison life for some forty years until they retired and sold it in 1996. A large folio-sized publication,

> *If we had nothing else to brag of, we might boast of the health of our town. With the exception of a few cases of the 'Grippe, which takes the gentlest hold possible of our people, in consideration, perhaps, of their general good health, we have had no sickness in Madison this season. With the purest atmosphere and best of water, and best of living, and best of society, why should we not enjoy the blessing of health?*
>
> The Madisonian
> March 29, 1879

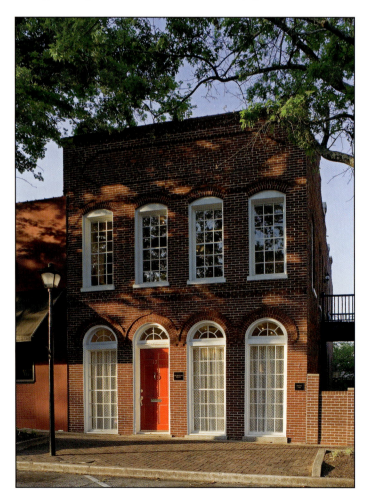

Offices of the Madisonian *on East Jefferson Street.*

the *Madisonian, Architecture and History*, which the Ponders began publishing for the annual spring tour of homes in the 1980s, was extremely helpful in research for the production of this bicentennial book. Adelaide Ponder wrote in the 1986 edition, "We are also including a brief history, 'Madison: A Town Proud of its Rich and Colorful Antebellum History,' compiled for the *Madisonian* in 1976 in hopes that some day a real history of Madison will be written."

In addition to its obvious but often overlooked contributions in building the architectural environment we now commemorate, the Black population of the town and county has produced accomplished and celebrated observers of this world as well, among them Benny and Raymond Andrews, who were two of ten Andrews siblings raised in Plainview, near Madison. Benny Andrews (1930–2006) was a painter, sculptor, printmaker, illustrator, teacher, and activist, whose works are in the collections of museums throughout the world. His brother Raymond (1934–91) was a widely acclaimed author whose life's work was recognized most recently by his 2009 induction into the Georgia Writers Hall of Fame.

Our book details the contributions to the society of this special place since its early days. Some of the other names that must be included are Carroll Hart, state archivist and historian; Caroline Candler Hunt, historian; Curtis Butler, Morgan County's first Black commissioner and a state leader of the NAACP; E. Roy Lambert Jr., lawyer, state senator, historic preservationist, and first chairman of the Madison Bi-Racial Committee; George Williams, community leader and second chairman of the Bi-Racial Committee; Robin Ferst, founder of the Ferst Foundation for Childhood Literacy; and Mrs. Kirby Smith (Susie) Anderson, former Morgan County Historian. We should also note four public officials who provided crucial leadership in the preservation movement, Luke Allgood, Barry Lurey, Bruce Gilbert, and Fred Perriman. Finally, there is Marshall "Woody" Williams, who retired from decades as an educator in Madison and Athens and now serves as county archivist and historian, despite having never received compensation for his many years in this capacity.

This tribute to Williams appeared on the first page of the spring 1992 issue of the *Madisonian, Architecture and History*: "If you're looking for Marshall 'Woody' Williams, walk over to the Old Morgan County Jail [restored to house the archives as a result of his efforts and now named in his honor]. Deep in the recesses of the cavernous building Williams can be found, searching and compiling, hovered over countless historical documents he has saved from sure destruction. . . . Through countless hours of work and through endless lobbying efforts with local governments, he has taken it upon himself to recognize and record Morgan's colorful historical documents."

Much of Madison's history is written on the headstones and monuments in the Madison Old Cemetery.

> *All that last night you wrote notes*
> *For the disposition of your manuscripts, books,*
> *When autumn had come gold and red*
> *Back to Georgia. You took the weave of age,*
> *Spread that uneven tapestry half across*
> *Your house in the woods. You came*
> *Past old lapses, memories of baseball,*
> *Funny-paper stories from the Thirties*
> *When you and Benny were only boys*
> *In Madison. I consecrate all the layers*
> *Of that last long evening before us.*
> *from "In Memory of Raymond Andrews"*
> *Philip Lee Williams, 2009*

County and Region, Towns, Tourism, Hard Labor Creek State Park, and Lake Oconee

Morgan County was established December 10, 1807, on the rich soil of the Georgia piedmont plateau frontier, all lands west of it being Creek Indian territory. In 1813, for example, Creek Indians attacked settlers in the area that became Rutledge and Hard Labor Creek State Park. The Creeks were allied with the British.

The county was named for General Daniel Morgan (1733–1802), American hero of the Revolutionary War Battle of Cowpens in South Carolina. Morgan County was created out of Baldwin County, as Milledgeville was being established as the new state capital and the county seat of Baldwin, about forty miles southwest. Athens, founded as the home of Franklin College (now the University of Georgia), is about thirty miles northeast of Madison, reached on U.S. 441 through Apalachee, one of the oldest communities in Morgan County. In 1807, of course, Atlanta, sixty miles northwest, was not yet even the railroad terminus that it would become in the 1840s.

As the county grew, the first town to develop was Buckhead, about seven miles east toward Greensboro, the county seat of Greene County, now home of Lake Oconee, a portion of which extends into Morgan. When the Georgia Railroad came to Madison in 1841, the next stop east was Buckhead. By that time Buckhead had a post office and a general store. Today, Buckhead, Georgia, calls itself, "the Real Buckhead," because of the large suburban area by that name in northwest Atlanta.

Rutledge, west of Madison, was an early Georgia Railroad stop, although it was not incorporated until December 13, 1871. Just northwest of Rutledge, towards Fairplay, is Hard Labor Creek State Park. This large park was carved out of the Georgia landscape in the early 1930s during President Franklin Roosevelt's New Deal by the Civilian Conservation Corps; it was a CCC camp until 1939, when it became a national park, which it remained until 1946 when the state took over. The park today is best known for its eighteen-hole public golf course. Its six thousand acres offer cottages, hiking, horseback riding and trails, and Lake Rutledge.

The Civilian Conservation Corps built Lake Rutledge as part of Hard Labor Creek Park.

Tourism had begun in Morgan County in the antebellum 1840s and '50s, with hotels and the like opened for tourists. Earlier there were inns and taverns to accommodate stagecoach passengers. By the time of Hard Labor Creek Park and automobiles, the area had become a major modern tourist destination, which it remains. Hard Labor Creek cuts through the park; it may have been named by slaves who tilled the fields in antebellum days. There is also a story that Native Americans called it Hard Labor because of the difficulty they had crossing it. Today the creek is considered a golf hazard, so its historic name, Hard Labor, continues to be appropriate! Easily accessed from Interstate 20, with Atlanta about fifty miles west, this state park has long been like a Madison country club.

A bustling community in the 1890s, Bostwick was incorporated in 1902. A cornerstone of the local economy at the time was a cotton gin owned by John Bostwick. Today, more than a century later, his great grandson John Bostwick is the mayor, and the cotton gin is the only one still operating in Morgan County. One of the few remaining old hotels, the Susie Agnes, is now the Bostwick City Hall.

One of the other Morgan County towns and villages that should be noted is Park's Mill, located on the banks of the Oconee River, dating from about as early as the foundation of the county. Early on, there were a grist mill and ferry. The Parks' family home served as a general store, tavern, and inn for travelers. Later, it had a post office. In 1981

Above: Bostwick.
Below: Park's mill on the Oconee River.

the Parks' house, one of the oldest in the county, was moved a few miles away from its original site because of the creation of Lake Oconee.

Pennington is named for the well-known Morgan County Pennington clan that founded multi-million-dollar Pennington Seed, Inc., in 1945. The unincorporated community is located on highway 83, about ten miles southwest of Madison. The village crossroads has had a general store and, since 1890, a Methodist church. The Pennington homeplace, built in 1847, is still owned by the family. Other notable communities in the county are Dorsey, Godfrey, High Shoals, and Swords.

Lake Oconee, east of Madison, has added new features to recreation, life, and tourism in Morgan County. One aspect has been to give the name Lake Country to this part of Georgia. Lake Oconee, the state's largest with nineteen thousand acres under water, is a Georgia Power Company reservoir dating from 1979 and fills a portion of the boundaries between Morgan, Greene, and Putnam counties. The area is known for its upscale residential developments like Reynolds Plantation, Harbor Club, Port Armor, and Great Waters, and as a world-class golf destination.

John Buck Swords house in Swords community.

Along the Morgan County shoreline, a group of Madisonians developed a retreat for weekend and summer recreation around a secluded cove. One of the families, the Ponders, rescued the John C. Wood house (c. 1818) from its derelict condition and moved it to the cove, where it was restored by Atlanta architect Norman D. Askins.

Without a doubt, Lake Oconee has added to life in Morgan County, Madison, and the region. In 2001 Madison was voted the best small town in America by *Travel Holiday Magazine*. That Morgan County "lays claim to the northernmost part of the lake," as another travel magazine put it, is one of the reasons for such a complimentary designation in the decade of the town and county's bicentennial years.

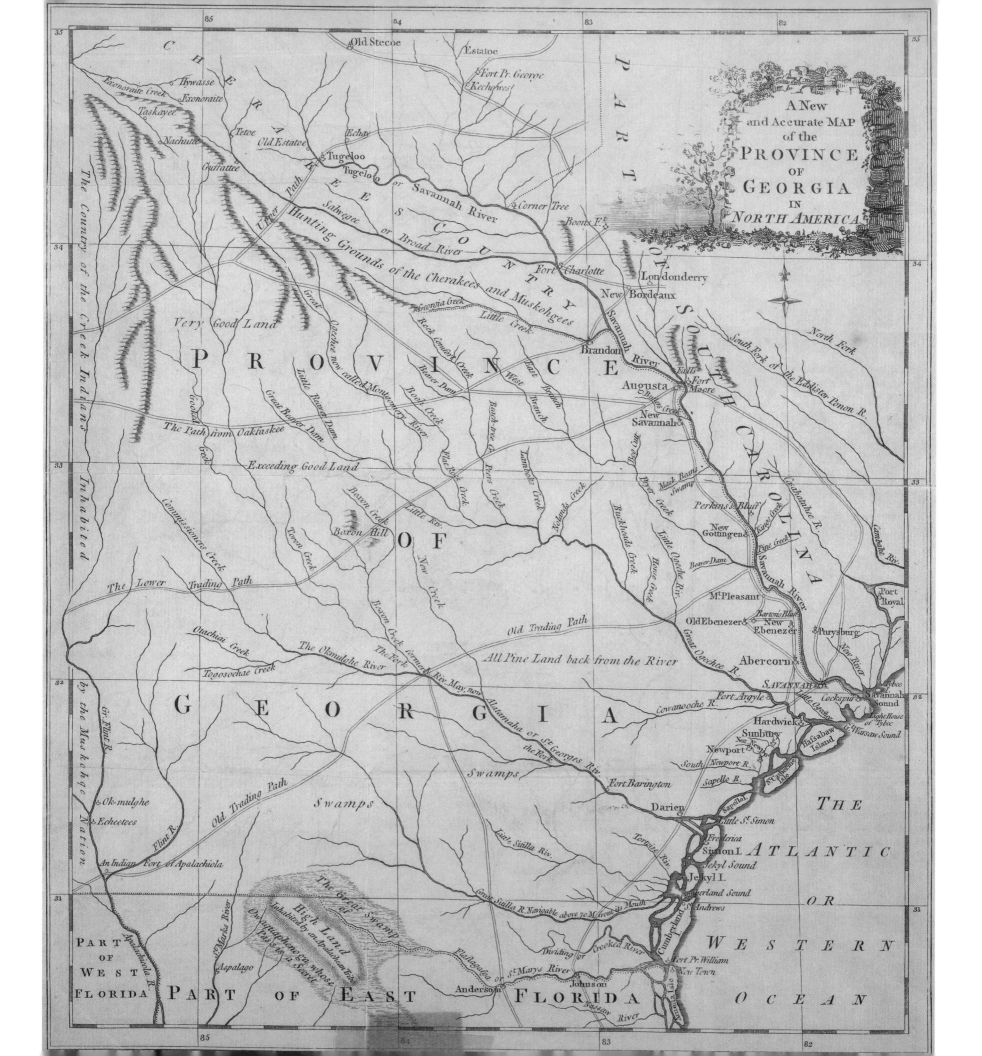

A Timeline of Historical Context for Madison and Morgan County

1540 Spanish explorer Hernando de Soto led an expedition deep into the territory that would become Georgia, encountering various Native American chiefdoms.

1733 February 12. James Edward Oglethorpe established the town of Savannah and began the settlement of Georgia, the thirteenth English colony. Oglethorpe negotiated the first cession of Creek lands to the Georgia colonists.

1751 January 1. The ban on slavery in colonial Georgia, which had been in place since 1735, was lifted by the English House of Commons.

1763 The Treaty of Augusta settled outstanding debts by the Creek and Cherokee to the British through cessions of land in the interior of Georgia.

1775 April 19. The first shots of the Revolutionary War were fired at Lexington, Massachusetts, as British troops fired into a crowd. Later that day local militia defeated British regulars at nearby Concord.

1781 January 17. Continental troops under the command of General Daniel Morgan dealt a humiliating defeat to British regulars led by Lieutenant Colonel Banastre Tarleton at the Battle of Cowpens in South Carolina. This surprising victory is considered a turning point in the Revolutionary War, coming just as the British command believed it would soon take control of the entire American South.

1783 April 15. A provisional peace was struck between England and the American Colonies at the Second Treaty of Paris.

1788 January 2. Georgia achieved statehood, becoming the fourth of the newly created United States of America.

1790 Total population of Georgia: 82,548;
Slave population: 29,264.

"We passed through her [the queen of Cutifachiqui] countrie an hundred leagues, in which, as we saw, she was much obeyed. For the Indians did all that she commanded them with great efficacie and diligence." [a Fidalgo of Elvas] After a march of seven days the province of Chelaque was reached. In this name, with but slight alteration, we recognize the land of the Cherokees.

from accounts of de Soto's expedition
in The History of Georgia
Charles C. Jones Jr.

Engraving of gold medal awarded to General Morgan.

Dear Sir—The troops I have the honor to command have gained a complete victory over a detachment from the British Army commanded by Lieut.-Col. Tarleton. It happened on the 17th inst., about sunrise, at a place called the Cowpens, near Pacolet River.

letter from General Daniel Morgan to
General Nathanael Greene
January 19, 1781

Opposite: Map of Georgia in 1779.

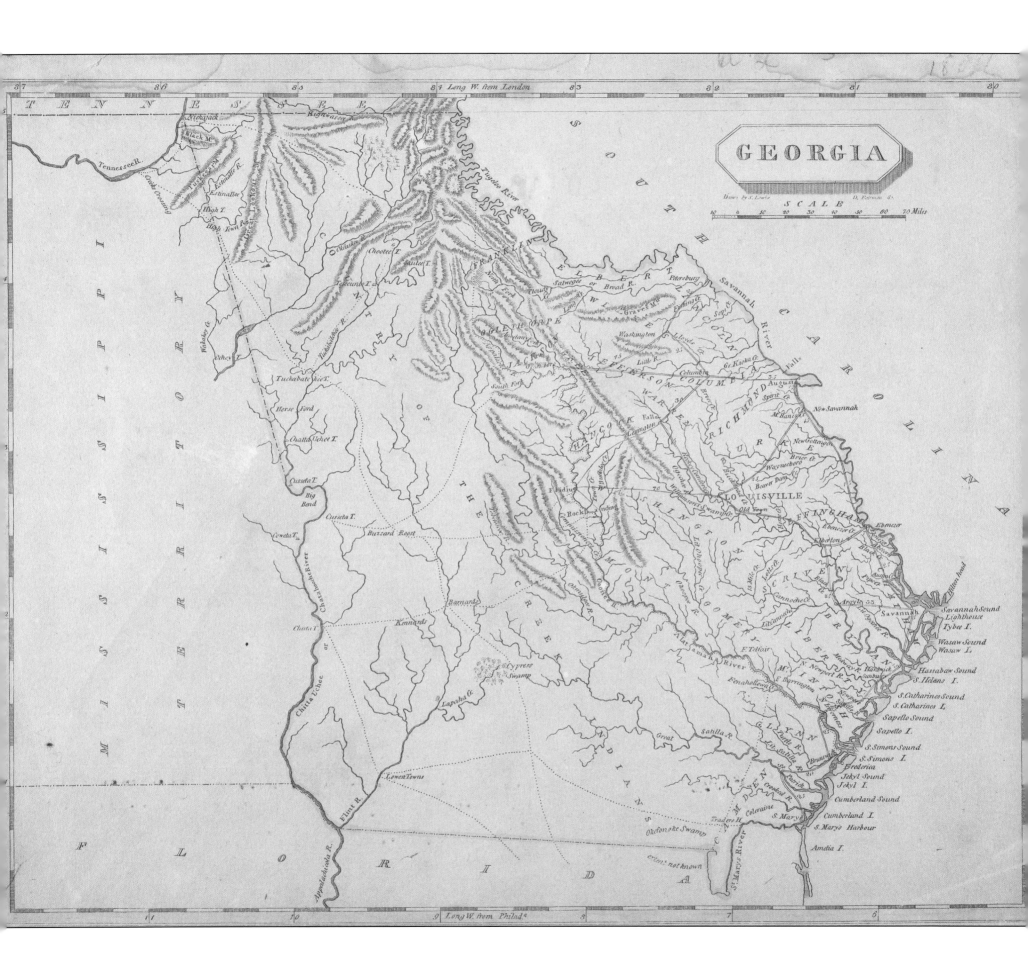

Opposite: Map of Georgia in 1805.
Above: Map of Georgia in 1818.

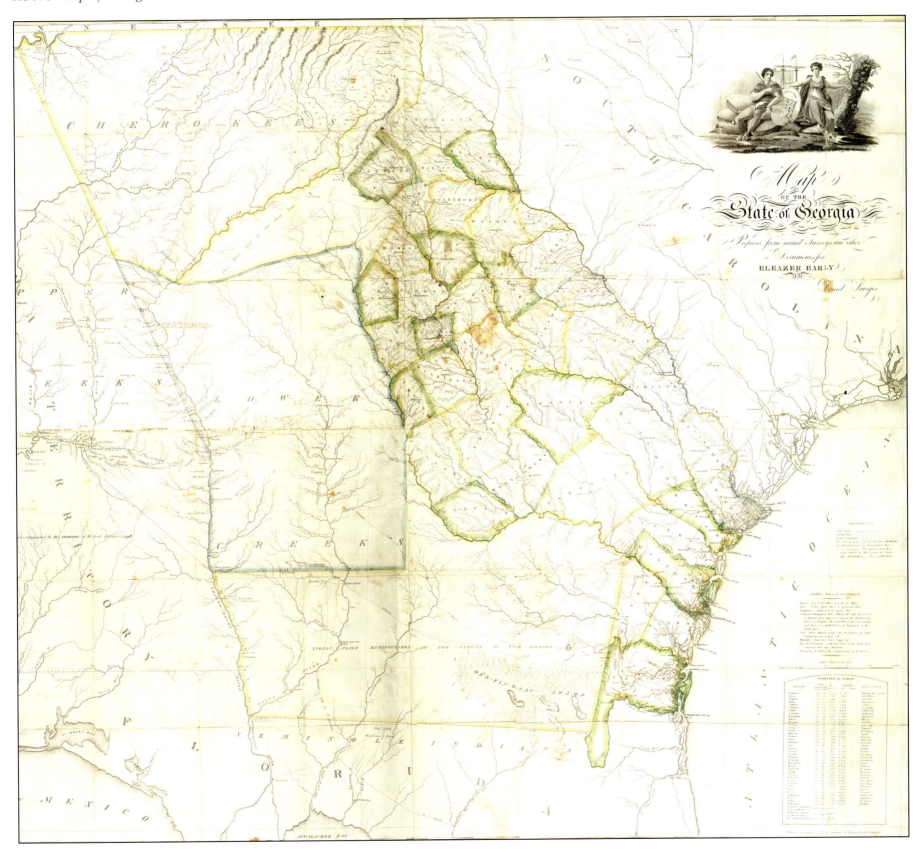

ARTICLES of a Convention made between Henry Dearborn, secretary of war, being specially authorized therefor by the President of the United States, and Oche Haujo, William M'Intosh, Tuskenehau Chapce, Tuskenehau, Enehau Thlucco, Checopeheke, Emantlau, chiefs and head men of the Creek nation of Indians, duly authorized and empowered by said nation. ART. I. The aforesaid chiefs and head men do hereby agree, in consideration of certain sums of money and goods to be paid to the said Creek nation by the government of the United States as hereafter stipulated, to cede and forever quit claim, and do, in behalf of their nation, hereby cede, relinquish, and forever quit claim unto the United States all right, title, and interest, which the said nation have or claim, in or unto a certain tract of land, situate between the rivers Oconee and Ocmulgee (except as hereinafter excepted) and bounded as follows. . . .

*from the Treaty of Washington
November 14, 1805*

General Daniel Morgan.

1793   Eli Whitney designed the first highly efficient cotton gin at Mulberry Grove Plantation near Savannah, setting the stage for the boom in cotton production and the need for large land holdings and a labor force to support it.

1800   Total population of Georgia: 162,686;
Slave population: 59,699.

1802   Treaty of Fort Wilkinson defined additional cessions of Creek land west of the Oconee and Apalachee rivers.

1804   December 12. The Georgia state legislature passed an act calling for the establishment of a town in the county of Baldwin, part of the newly ceded territory. The town would be called Milledgeville, and three years later it would be the state capital, Georgia's fourth.

1805   The first Treaty of Washington marked the agreement for further cession of the remaining land between the Oconee and Ocmulgee rivers. This treaty also called for the establishment of a road through the territory, which followed a Creek trading path and became a part of the federal road from Philadelphia to Charleston to New Orleans.

1807   December 10. The state legislature passed an act creating six new counties, the first of which was Morgan County, named for Revolutionary War General Daniel Morgan, who had inspired his forces, many of whom were from Georgia, to victory at the Battle of Cowpens. Land lotteries in 1805 and 1807, intended to encourage settlement by individual yeoman farmers, accounted for the distribution of most of the new county in parcels of 202.5 acres each.

Park's Bridge was built over the Oconee River in eastern Morgan County. A small community was located there, variously known as Park's Mill, Park's Bridge, Park's Ferry, or Riverside.

Methodists organized in the new county. The Apalachee Circuit was served by Reverend Lovic Pierce. The following year a wooden church was erected in the area that would become Madison.

1808	December 7. Electoral college delegates cast their individual ballots to elect the fourth president of the United States in a contest between James Madison and Charles C. Pinckney.

December 22. The seat of government for Morgan County was to be named Madison, after James Madison, presumptive president-elect, coauthor of the *Federalist Papers*, a framer of the Bill of Rights, and often referred to as the Father of the Constitution.

1809	January 19. An entrepreneurial group assembled land from lottery lots 35 and 36; surveyor Lewis McLane devised a town plan based on the Washington, Georgia, example, with a central public square defined by four principal streets, which would be named Monroe (now Main), Jefferson, Hancock, and Washington. Each of the original forty-eight lots measured 100 by 200 feet.

February 8. Congress announced James Madison had won the electoral college vote 122 to 44 and would be the new president.

February 23. The first twenty-six lots in Madison, comprising approximately thirty-five acres, were sold. Reuben Rogers purchased lot number six in the new town and began construction of his house.

March 4. James Madison was inaugurated.

December 12. Madison was officially incorporated, becoming the first new town in any state of the young Union to be named to honor President James Madison. A board of commissioners was appointed, and Madison was made the county seat.

Born in 1809: Abraham Lincoln and Charles Darwin (both on February 12), Louis Braille, Edgar Allen Poe, Cyrus McCormick, Alfred Lord Tennyson, Oliver Wendell Holmes Sr., Fanny Kemble, and Kit Carson.

1810	Total population of Morgan County: 8,369;
Slave population: 2,418;
Total population of Georgia: 252,433;
Slave population: 105,218.

President James Madison.

Reuben Rogers house (1809–10), photographed in 1973.

Above: Johnston family plot in the Madison Cemetery. Below: Monument to Benjamin Braswell in the Madison town square.

1810 Construction began on the Morgan County Courthouse and continued for fifteen years before completion.

A private school for boys was established as the Morgan County Academy and was later incorporated as the Madison Academy. By 1820, 745 students studied Greek, Latin, geography, grammar, arithmetic, reading, and writing in a two-story brick building known as the Male Academy. It burned in 1824.

1811 Dr. William Johnston built a frame house on West Central Avenue, which was purchased in 1832 by Dr. Elijah Evan Jones, who subsequently remodeled it in the Greek Revival style. Now facing South Main Street and known as Heritage Hall, it is a house museum operated by the Morgan County Historical Society. Dr. Johnston's infant daughter, Elizabeth, died July 11, 1811, and was buried in the Old Madison Cemetery, where her grave is the earliest marked interment.

1812 June 12. The United States declared war with England. The War of 1812 had begun.

1813 Creek Indians, who supported the British, attacked white settlers near what is now the town of Rutledge in western Morgan County.

1814 December 24. The Treaty of Ghent ended the War of 1812, but the fighting went on for two more months.

1817 Benjamin Braswell established a trust for the education of children of destitute widows.

Buckhead was settled about this time in eastern Morgan County and by 1823 had its own post office. Although it attempted incorporation in 1887, it was not formally incorporated until 1908. Buckhead has a long history of commerce, certainly enhanced when the Georgia Railroad line was completed through town in the 1830s.

1819 December 22. John Colby was given the right to operate a line of stages (carriages) from Madison to Hancock County twice weekly.

1820 Total population of Morgan County: 13,520;
    Slave population: 6,045;
    Total population of Georgia: 340,989;
    Slave population: 149,656.

1821 Around this time many small, wooden houses were built in Madison, including a home for Judge William Stokes's family on Old Post Road. Now known as the Stokes-McHenry house, it is among several in town to have a very early, simple core, which has been expanded and renovated over succeeding generations.

1822 December 23. Madison town limits were extended to include roughly "all land within one-half mile of the public square." This radius created a circular town, characteristic of early town planning in Georgia.

Engraving of a fledgling Georgia town around 1820.

1824 Three one-acre lots in Madison's town commons were laid out for Baptist, Methodist, and Presbyterian congregations.

I repaired promptly to my work and found among my flock in Madison a kindly welcome. Madison was then a thrifty little village, and there was a respectable society worshiping in what we considered a pretty decent wooden church.
    Methodist Reverend James Osgood Andrew, 1829

1830 Total population of Morgan County: 12,046;
    Slave population: 6,820;
    Total population of Georgia: 516,823;
    Slave population: 217,531.

Snow Hill.

Lancelot Johnston, a Madison planter and inventor, received a patent for his cottonseed huller. He found that the oil he extracted from cottonseeds could be mixed with white lead to form a substance he used to paint his house entirely white, including the roof. He named the house Snow Hill.

1833 The Georgia Railroad Company was incorporated to build a line from Augusta to the Chattahoochee River with branches to Athens, Madison, and Eatonton.

On a Morgan County plantation originated an economic process which today underlies one of the greatest industrial activities of the world—the manufacture of cotton seed oil. . . . The first successful effort ever made to extract oil from cotton seed was made by Launcelot Johnstone [sic], Esq., within a quarter mile from the courthouse in Madison.
    Lucian Lamar Knight
    Georgia and Georgians, *1917*

Above: Hilltop (1838).
Below: Bonar Hall (1839) with original portico.

Madison Presbyterian Church (1842).

1834 Baptists organized as a congregation and built a meeting house on Academy Street.

1836 Farmers Hardware Company opened on the southeast corner of First and Jefferson streets. Now located on the Atlanta Highway, it is Madison's oldest surviving business.

1838 The Edmund Walker Town House was built on Academy Street.

About this time Thomas J. Burney began building for his bride a Plantation Plain house on North Main Street. Called Hilltop, its functional, vernacular plan was embellished with Greek Revival details.

1839 Construction began on Bonar Hall, a neoclassical brick home built by John Byne Walker. Renovations, including Victorian porches, were made in the 1880s by the Broughton family.

1840 Total population of Morgan County: 9,121;
    Slave population: 5,646;
    Total population of Georgia: 691,392;
    Slave population: 280,944.

A Georgia Railroad line opened to Madison, enabling freight from the cotton-rich piedmont to be shipped to Augusta and on to Savannah.

1841 The Georgia Railroad Terminal was built. It was significantly damaged by Union torches in 1864 and was rebuilt with alterations.

The American Hotel opened on the corner of North Main and West Jefferson streets.

1842 The Madison Presbyterian Church was built on South Main Street.

Southern Miscellany, a Madison weekly newspaper, was first published by C. R. Hanleiter. It ceased publication in 1847 and was succeeded that year by the *Madison Family Visitor*, which became the *Georgia Weekly Visitor* in 1859 and was published until 1861.

1843 Humorous stories of Madison and Morgan County written by William Tappan Thompson, editor of the *Southern Miscellany*, were collected in the book *Major Jones's Courtship*.

1844 The Methodist congregation built a brick Gothic Revival church on Academy Street. It is now the Episcopal Church of the Advent.

The first Morgan County Courthouse, built c. 1810 in the town square, burned.

1845 The second Morgan County Courthouse was built in the town square on the site of the first courthouse.

The hamlet of Marthasville, about sixty miles west of Madison, was renamed Atlanta. The westward extension of the Georgia Rail Road was completed between Madison and Atlanta, supplanting the branch to Athens as the main line.

1848 Thurleston, originally a five-room Plantation Plain house built in 1818 and moved to its present site in 1841, was enlarged and remodeled by Dr. Elijah Jones with architectural planning by Benjamin Peeples.

1849 Madison town limits were expanded to a one-mile radius from the town square.

Madison is the county town, situated on the ridge which divides the waters of Sugar and Hard Labour creeks, surrounded by beautiful and fertile country. The court-house is a spacious brick building and the jail is constructed of granite. In the town are three churches, Methodist, Presbyterian, and Baptist, all neat edifices; two hotels, eight dry goods stores, one printing office, &c.
                                                            George White, 1849

1850 Total population of Morgan County: 10,744;
Slave population: 7,094;
Total population of Georgia: 906,185;
Slave population: 381,682.

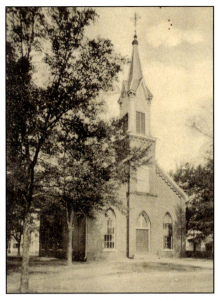

Above: Methodist Church (1844), now Episcopal Church of the Advent. Below: Morgan County Courthouse (1845).

Thurleston (1818), enlarged, and remodeled in 1848.

Above, left to right: Martin-Baldwin-Weaver house (1850); Carter-Newton house (1850); Massey-Tipton-Bracewell house (c. 1854).
Below, top to bottom: Billups-Tuell house (c. 1853); Georgia Female College (now the Baldwin-Williford-Ruffin house) (c. 1850); Honeymoon (1851).

1850 A building boom of significant houses took place in and around Madison in the years immediately surrounding the midpoint of the century. Notable houses in the Greek Revival style were built for the Martin, Massey, Carter, and Billups families.

All in all, the Georgia Greek Revival house displays great variety— from pomposity and stolid dignity to graceful improvisation and even whimsicality, from parvenu grossness to delicate romanticism, . . . The Greek Revival was not limited to the great two-story mansion in which it registered its boldest achievements. In a few of the towns of Piedmont . . . the classical style was attempted with charming results in the one-story cottage.

Wilbur Zelinsky, 1954

Madison began its rise as an early center of advanced education for women. The Georgia Female College, first incorporated as the Madison Collegiate Institute, was founded by the Baptists. The Methodists founded the Madison Female College. Both had handsome brick classroom buildings which later served as hospitals during the Civil War and were eventually lost to fire. The first building at the Georgia Female College is now a private residence, the Baldwin-Williford-Ruffin house on South Main Street.

1851 Charles Mallory Irwin built a Greek Revival home south of Madison's downtown on the Eatonton Road. One of the town's signature antebellum, white-columned houses, it was named Honeymoon in the 1930s by Mrs. Peter Godfrey.

1851 Boxwood, a grand home built in the transition from Greek Revival to the Italianate style, was begun by the Kolb family on Old Post Road.

The Kolb house . . . boasts a rare possession, twin box gardens. . . . Though a town house, the service yard which in this case lies to the side, contains not only carriage house and stables but a cow barn as well, and is flanked on one side by a vegetable garden and orchard. . . .
Garden History of Georgia, *1933*

City Market was built on the town square and served as a centralized public vending and trading venue until the late 1880s.

1858 The Madison Baptist Church was constructed on South Main Street. The portico was added in 1917.

1859 The Home Guards, a volunteer corps, was incorporated with members coming from the city and county.

1860 Total population of Morgan County: 9,997;
Slave population: 7,006;
Total population of Georgia: 1,057,286;
Slave population: 462,198.

November 6. Abraham Lincoln was elected President.

December 20. South Carolina seceded from the Union.

1861 January 19. Delegates to a special Georgia convention on secession voted 208–89 in favor, joining South Carolina, Florida, Mississippi, and Alabama. Soon Louisiana and Texas would follow, and by the end of May, Virginia, Arkansas, Tennessee, and North Carolina would also secede.

February 9. The Confederate States of America was formed.

April 12. Confederate forces under General P. G. T. Beauregard fired on federal Fort Sumter at Charleston. The War between the States (also known as the Civil War) had begun.

Top: Boxwood (1851), the Italianate elevation. Above: The garden plan was drawn for the Garden History of Georgia.

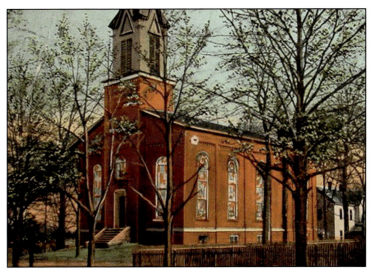

Madison Baptist Church (1858) before alterations were made to the façade and steeple.

The many fine residences were built in the same manner of all the better class of Southern homes, extensive piazzas in front with tall fluted columns reaching almost to the top of the house. . . . We felt that the people of these seemingly prosperous cities were more responsible for the war than those whose farms we had overrun since leaving Atlanta. We wanted to impress on them some idea of the power and magnitude of the Army they so hated and despised.

Rice Bull, Union Army, November 18, 1864

Union troops entering Madison, November 18, 1864, from an engraving in Harper's Weekly.

[A]lthough opposed . . . to your forces, it will . . . be a satisfaction to you to know that he fell at the head of his Brigade—honorably battling for the cause he thought just and righteous. . . . Mrs Helm is crushed by the blow—almost broken hearted—and desires to return to her Mother and friends in Kentucky. . . . Mrs Helm desires to be affectionately remembered to her sister.

letter from E.M. Bruce to President Abraham Lincoln, 1863

1861 July 31. The Panola Guards, composed mainly of Madison residents, left for Richmond, Virginia. During the course of the war, most of Madison's able men would be in armed service to the Confederacy, leaving the town to be managed by aged or ill white men. White women and children and black slaves would form the fabric of the town and the surrounding countryside for the next four years.

1863 January 1. President Lincoln's Emancipation Proclamation was to take effect, stating that "all persons held as slaves within any State or designated part of a State, the people whereof shall then be in rebellion against the United States, shall be then, thenceforward, and forever free."

September. After her husband was killed at Chicamauga serving as a general for the Confederacy, Emilie Todd Helm, half-sister to Mary Todd Lincoln, spent time in Madison at the home of E. M. Bruce, who wrote a letter on her behalf seeking a pass to her mother's home in Kentucky. She ultimately went on to the White House and stayed with the Lincolns for awhile.

1864 November 15. Union forces, numbering 62,000 men under General William T. Sherman, left Atlanta to begin the infamous March to the Sea.

November 16. Union troops destroyed the Rutledge train depot, water tank, and most of the railroad tracks.

November 18–19. Union troops occupied Madison, destroying railroad tracks and the Madison rail depot.

November 20. Federal raiders destroyed Park's Mill and Ferry.

1865 January 31. The Thirteenth Amendment to the Constitution, which abolished slavery, was approved by Congress.

March 3. The Bureau of Refugees, Freedmen, and Abandoned Lands was established under the War Department. The Freedmen's Bureau, as it became called, was created to assist freed slaves and displaced persons within the southern states and the District of Columbia under Reconstruction authority.

1865   April 9. General Robert E. Lee surrendered to Union General Ulysses S. Grant at Appomattox Court House, effectively ending the War between the States.

1866   Calvary Baptist Church, Madison's first independent Black congregation, was established. It worshiped in the frame church built in 1834 by Baptists on Academy Street.

1867   The Freedmen's Bureau purchased land on Hill Street for the first school for African Americans in Madison. The bureau first rented then bought the frame church on Academy Street, disassembled it, and moved it to the Hill Street property where it was repurposed and used as the Madison Freedmen's School. When Calvary Baptist built its new church on Academy Street in 1883, a new congregation formed in the Hill Street church, which became known as Clark's Chapel after Reverend Allen Clark.

1868   Joshua Hill, an ardent anti-secessionist from Madison, whose son died in service to the Confederate Army, was elected to the United States Senate.

The state capitol was moved from Milledgeville to Atlanta.

1869   A devastating fire destroyed forty-two structures in downtown Madison with the courthouse and hotel on the square surviving. The town responded quickly, however, erecting brick, "fire-proof" buildings; the general appearance of the commercial district in 2009 was shaped during the construction recovery in the 1870s and '80s.

1870   Population of Madison: 1,389;
   Population of Morgan County: 10,696;
   Population of Georgia: 1,184,109.

Weekly newspapers of the era: the *Georgia Home Journal* (1871–75), later known as the *Madison Home Journal* (1876–78), and then as the *Madisonian* (1879–2001). Other local publications included the *Southern Farmer and Stock Journal* (c. 1867–76), the *Advertiser* (c. 1887–1905), and an independent Black paper, the *Gleaner* (c. 1899–1901).

Clark's Chapel was originally the Baptist church building on Academy. It was then moved and reconstructed on Hill Street.

The Atkinson house (c. 1854) on Wellington Road was renovated by adding Gothic Revival gables in the late 1860s.

A brick commercial building (168 South Main Street) built after the fire of 1869.

The Albert Douglas house (c. 1870) at 614 South Main.

Above: Saint Paul African Methodist Episcopal Church (1882). Below: Calvary Baptist Church (1883).

1871 Rutledge, in western Morgan County, was incorporated. Developed around a terminus and roundhouse for the Georgia Railroad in the 1840s, Rutledge was named for Hezekiah Rutledge, whose widow deeded land for the railroad right-of-way.

The Morgan County Board of Education was elected.

The Madison Hotel was built. A two-story building on the east side of the town square, it burned in 1891.

1876 February 18. The municipal charter was amended so the Town of Madison became the City of Madison with leadership officially designated in a mayor and board of aldermen instead of a president and board of commissioners.

1879 The local newspaper was published as the *Madisonian*.

Godfrey's Warehouse, a feed mill and retail operation, began serving Madison and the surrounding environs. Still owned and operated by Godfrey descendants, it is the oldest Madison business still run by the same family.

1880 Population of Madison: 1,974;
    Population of Morgan County: 14,032;
    Morgan County cotton production: 7,358 bales;
    Population of Georgia: 1,542,180.

Land for the Madison New Cemetery was acquired by the city across the railroad from the Madison Old Cemetery.

1882 The Saint Paul African Methodist Episcopal Church was completed on Fifth Street in the Canaan neighborhood on land purchased in 1871.

1883 The Calvary Baptist Church was completed on Academy Street on land acquired in 1872 from the Madison Baptist Church.

The Hunter house, an elaborately decorated rendition of the Second Empire Style, was built on South Main Street.

Madison will double her population in ten years, and all the croakers of christendom can not prevent it. Our little city has a future, and it is well that she is preparing for it.
  The Madisonian, *June 8, 1884*

Cotton production boomed between 1880 and 1890.

1885 The Covington and Macon Railroad Company (now the Central of Georgia Railroad) was chartered to build a line from Macon to Covington, but it was altered to run through Madison to Athens. The line reached Madison in 1888 and was completed in 1889.

Bonar Hall, built in 1839–40, was renovated by removing the original one-story portico and replacing it with a two-story Victorian porch.

Enlargement and remodeling began on the Martin Richter house, now called Dovecote, on Main Street. This period saw the remodeling of several houses around town with Victorian embellishments manufactured at the Madison Variety Works, which provided the decorative trim for many of the houses, small and large, built in Madison at this time.

1887 The Madison City Hall with a fire house and jail was built on East Jefferson Street. Some city offices, including the mayor's and city clerk's, were moved into the old courthouse in 1909 and remained there until it burned in 1916. The 1887 city hall now houses the Madison Chamber of Commerce and Visitors' Center but retains the fire pole, bars on the rear calaboose windows, and the construction date near the cupola.

Above: Bonar Hall (1839–40) after renovations in 1885. Below: The antebellum Martin Richter house on South Main Street after remodeling in the 1880s and '90s.

1888 Bank of Madison, the community's oldest financial institution, was granted a charter by the Georgia General Assembly.

1890 Population of Madison: 2,131;
 Population of Morgan County: 16,041;
 Morgan County cotton production: 19,300 bales;
 Population of Georgia: 1,837,353.

Top, left to right: Peter Walton house (c. 1889) on South Main Street; the Madison Hotel fire in 1891; Morgan County Jail (1892). Below: Turnell-Butler Hotel (c. 1892).

Above: South Main Street School (1895).

1891 Madison installed an electrical system, operating a generating plant until the franchise was sold to Georgia Power in 1929.

The Madison Hotel on Hancock Street burned.

Born in slavery, Adeline Rose built a frame, one-story house near the railroad in Madison for her family. In 1996 her home was acquired and moved to East Jefferson Street, where it as renovated by the City of Madison and is operated as a companion house museum to the Rogers House.

1892 The county jail, a brick building with a pyramidal-roof tower, was erected on Hancock. It is now the repository of the Morgan County Archives.

The Turnell-Butler Hotel (later the Hotel Turnell, the Hotel Madison, and then the New Morgan Hotel) was built on Hancock Street and operated for the first year by Oliver Hardy (father of comedian Oliver Hardy).

1893 The Madison Garden Club, renamed La Flora around 1932, was founded. It was the second oldest garden club in Georgia, after the Athens Ladies Garden Club (1891), the oldest garden club in the United States.

1895 The City of Madison built its first public schools based on the racially separate system dictated by the state government: the South Main Street School (the Madison Graded School), a large brick building in the Romanesque Revival style, and the Burney Street School, an impressive, turret-topped frame structure in the Canaan area.

Above, left to right: Trammell house (1897–98), now Oak House; Porter-Fitzpatrick house (c. 1850), renovated in 1901; Poullain Heights (1905). Below: The antebellum Joshua Hill house (remodeled in 1917).

1900 Population of Madison: 1,992;

Population of Morgan County: 15,813;

Population of Georgia: 2,216,331

The turn of the century saw the building and remodeling of many significant houses in Madison in the Neoclassical Revival style. Among these are the Trammell (1898), the Porter-Fitzpatrick (remodeled in 1901), Poullain Heights (1905), and the Joshua Hill house (remodeled in 1917).

Weekly newspapers of the era: the *Madisonian* (1879–2001) absorbed the *Morgan County News* (1947–54) and was published for more than 130 years; the *Morgan County Citizen* took over the *Madisonian* (1997–2009).

1902 The town of Bostwick was incorporated. Bostwick owed much of its economic energy to cotton production, which continues today. A cotton gin owned by John Bostwick was among several vital businesses, including a bank, a hotel, a grist mill, and a second gin.

1904 Fairview, a private cemetery, was opened across the railroad and to the south of the Madison Old Cemetery.

1905 Construction began on a larger Morgan County Courthouse on the northeast corner of Jefferson and Hancock streets. It was completed in 1907.

The third Morgan County Courthouse was built on East Jefferson, facing the town square. The cupola of the 1887 Madison City Hall is to the left.

Above: Breaking ground for the Madison water works plant. In the background is the electric plant. Below: Heritage Hall (c. 1811 and c. 1830–40), at the time it was used as the Travelers Inn.

Below: The Eighth District Agricultural and Mechanical School campus, established in 1908.

Fourth of July Parade, 1905. Madison has a long history of civic events.

1907 The town of Apalachee, one of the oldest communities in Morgan County, with settlers in the area before 1820, was incorporated.

1908 Madison installed a public water works system.

The Eighth District Agricultural and Mechanical School was established on three hundred acres in the eastern part of Madison.

The house now known as Heritage Hall was moved from West Central Avenue to face South Main Street. Owned by the Turnell family, it was later converted from a residence into the Travelers Inn.

The Cooke Fountain was installed downtown. Later moved to the courthouse square, it was removed when the post office was built.

1909 Morgan County voted countywide taxation to support public schools.

The community of Swords was incorporated. Swords was the

home of the J. B. Swords Distillery, which developed near the early Morgan County community of Blue Springs, named for the natural spring water used in the production of Swords whiskey.

1910 Population of Madison: 2,412;
Population of Morgan County: 19,717;
Morgan County cotton production: 30,000 bales;
Population of Georgia: 2,609,121.

1913–14 The National Auto Trails movement promoted the improvement and use of recognized interstate roadways connecting state roads under single identifiable names. Madison was on two of the named routes, the east-west Jefferson Davis National Highway and a loop of the north-south Dixie Highway.

1914 June 28. Archduke Franz Ferdinand was assassinated in Sarajevo, Bosnia, precipitating the start of World War I.

The First United Methodist Church was built on West Central Avenue and South Main Street. It is a domed, neoclassical design across South Main from the Baptist and Presbyterian churches.

1916 The second courthouse building (1845) burned.

1916 Hill Park was developed on land given by Belle Hill Knight. The Confederate monument was moved in 1955 from downtown to the park.

1917 April. The United States entered World War I.

Below: Prairie style Douglas-McDowell house (1917).

Above: Cornelius Vason house (c.1910). Below: A circus parade about 1912 around the town square.

Above: First United Methodist Church (1914). Below: Cooke Fountain and Confederate Monument.

The Godfrey-Hunt house (1875) was remodeled in 1922 by architect Neel Reid.

Dedication of the World War I memorial with the statue "The Spirit of the American Doughboy."

1918 November 11. Armistice was signed, ending fighting in World War I.

1919 Morgan County cotton production: 36,197 bales.
Cotton prices peaked at 35 cents per pound.

1920 Population of Madison: 2,348;
Population of Morgan County: 20,143;
Population of Georgia: 2,895,832;
Cotton prices dropped to 17 cents per pound.

Main Street was paved, followed by the paving of ten miles of road between Madison and Rutledge and West Jefferson Street between the town square and the railroad depot.

The boll weevil began to wreak havoc on cotton production throughout the South.

1922 October 3. Rebecca Latimer Felton, a graduate of the Madison Female College, was appointed to the United States Senate. She was the first woman to serve as a United States Senator; her one-day term was the shortest in Senate history; and, at 87, she was the oldest person to serve in the Senate.

1924 Morgan County cotton production: 5,712 bales.

1930 Population of Madison: 1,966;
Population of Morgan County: 12,488;
Population of Georgia: 2,908,506.

May 4. A dedication ceremony was held in front of the courthouse for the unveiling of a statue honoring "the boys from Morgan County, Georgia, who fought in the World War April 6, 1917–November 11, 1918." The statue, "The Spirit of the American Doughboy," by sculptor Ernest Moore "Dick" Viquesney, is one of thirty-five known Viquesney doughboys at courthouses in the United States and was erected by the Henry Walton Chapter of the Daughters of the American Revolution.

The New Morgan Hotel burned.

1931   Construction began on the post office, a Colonial Revival building on the town square replacing the town park.

1934   Units of the Civilian Conservation Corps (CCC), created as part of President Franklin Roosevelt's New Deal economic recovery plan, began construction of a recreation area around Hard Labor Creek north of Rutledge. The CCC men cleared land, built roads, erected dams to create Lake Rutledge, and built camps, offices, and houses. Now Hard Labor Creek State Park, this was a national park until 1946, when it was turned over to the state for operation.

1939   A new Madison City Hall was completed in the Colonial Revival style on North Main Street. It also housed the jail and the firehouse.

September 1. Germany invaded Poland. Two days later Britain and France declared war on Germany. World War II had been set in motion.

Morgan County cotton production: 10,650 bales. In the 1930s new strains of insect-resistant cotton were developed.

The Episcopal Chapel, built in 1842 inside Madison Old Cemetery, was razed.

1940   Population of Madison: 2,046;
            Population of Morgan County: 12,713;
            Population of Georgia: 3,123,723.

1941   December 7. Japanese planes bombed the U.S. Pacific Fleet at anchor in Pearl Harbor. The following day the United States and Britain declared war on Japan.

December 11. Germany and Italy declared war on the United States. America had officially entered the war in Europe.

1945   May 7, Germany surrendered; August 14, Japan surrendered. World War II had ended.

Pennington Seed Company was founded in Madison.

Above: Madison Post Office (1931–32).
Below: CCC camp at Rutledge.

Madison City Hall (1939).

*Street scenes around the town square, from the early 1900s to the late 1940s.
Above left: Looking east on Jefferson Street toward the courthouse and Morgan Hotel. Above right: Main Street from Jefferson.*

Above: Miss Kittie Newton in her garden. Below: Parade commemorating Morgan County's 1957 sesquicentennial.

1945 The National Council of State Garden Clubs began a nationwide highway beautification program to pay tribute to World War II veterans. The Blue Star Highway system today recognizes U.S. veterans of all wars, and in Madison there is a Blue Star marker at the Boxwood Garden Club's Memorial Garden at highways 441 and 83 South.

1948 Morgan County consolidated the county high schools for white students into one school in Madison.

1950 Population of Madison: 2,489;
        Population of Morgan County: 11,899;
        Population of Georgia: 3,444,578.

The first Madison tour of homes took place, initiated by Miss Kittie Newton, owner of Boxwood and president of La Flora Garden Club.

Miss Kitty [sic] Newton . . . will try to make arrangements for interested visitors to see some of the homes listed here—at the owners' convenience of course. If she is away, the chamber of commerce, in the city hall, has agreed to help you in her place.
    A Guide to Early American Homes: South, *1956*

1957 September 14–20. Morgan County celebrated the 150th anniversary of its founding in 1807 with a week-long calendar of events in the local

Street scenes around the town square in the 1950s and '60s.
Above left: Looking north on Main Street. Above right: East Washington Street.

communities, including parades, balls, house and farm tours, a booklet on the county's history, and a pageant entitled Panorama of Progress.

1959   Morgan Memorial Hospital opened.

1960   Population of Madison: 2,680;
           Population of Morgan County: 10,280;
           Population of Georgia: 3,943,116.

Walton Park was developed on South Main Street through a bequest of Susan R. Manley.

1962   Construction began on Interstate Highway 20 in west Texas. The highway reached Madison in 1969.

1964   Madison appointed an ad hoc bi-racial committee, codified in ordinance the following year, to address integration issues and prevent conflict. Roy Lambert chaired the committee for the first three years and was succeeded by George C. Williams.

1970   Population of Madison: 2,890;
           Population of Morgan County: 9,904;
           Population of Georgia: 4,587,930.

August 25. The first day of classes began under the fully integrated school system of Morgan County. Following two years of voluntary integration, the last separate classes graduated in the spring from Pearl High School and Morgan County High.

The Morgan County Landmarks Society was founded.

The Morgan County 4-H Club ceded land for the development of Wellington Park.

1974   The Madison Historic District was listed in the National Register of Historic Places. It was later enlarged to encompass approximately 5,500 acres and more than three hundred buildings and sites.

1975   The City of Madison and the Morgan County Board of Education began a cooperative process that resulted in the development of Municipal Park.

1976   Georgia native James Earl "Jimmy" Carter Jr. was elected the thirty-ninth president of the United States.

The Madison-Morgan County Cultural Center opened in the renovated graded school building.

Scenes around the town square in December 2008.

1978 Cox-Elliot Memorial Park was developed following a Cox family bequest honoring Marilyn Cox Elliott.

1980 Population of Madison: 3,173;
Population of Morgan County: 11,572;
Population of Georgia: 5,463,605.

Georgia was selected as one of six states to be participants in the National Trust for Historic Preservation's fledgling National Main Street program, and Madison was one of the first cities in Georgia to be named a National Main Street community.

1987 The City Council passed an ordinance calling for the creation of a historic preservation commission.

1989 The Madison Historic District was created.

1990 Population of Madison: 3,447;
Population of Morgan County: 12,883;
Population of Georgia: 6,478,216.

1992 The City of Madison established the Rogers House Museum on the original town lot where Reuben Rogers built his house in 1809–10. The city negotiated the retention of the house by Morgan County, restored and interpreted the house and its grounds, and provided annual funding for its continued operation by a local nonprofit organization.

1993 The Morgan County African-American Museum opened in the c. 1900 home of John Wesley Moore. The house was moved to its present site on Academy Street for the museum.

1995 The City of Madison acquired land for Washington Park.

1996 The Atkinson family donated the "Point" for the city to develop Atkinson Park, honoring Grady and Camille Atkinson.

1999 The Ferst Foundation for Childhood Literacy started in Madison as a statewide initiative. As of 2009 it had distributed more than one million books to eighty thousand children in sixty-seven Georgia counties.

The National Arbor Day Foundation awarded Madison the designation of Tree City USA.

2000 Population of Madison: 3,636;
Population of Morgan County: 15,547;
Population of Georgia: 8,186,453.

Above: Madison Town Park and the James Madison Inn. Below left: The original Cooke Fountain (1908). Below right: A recasting of the Cooke Fountain was installed in the Town Park in 2009.

2002–3 Bostwick, Buckhead, and Rutledge historic districts were listed in the National Register of Historic Places, among them totaling 3,680 acres and more than two hundred historic buildings, structures, and sites.

2003 The final parcel of land was acquired for Round Bowl Spring Park, a linear native species park.

2004 Madison was honored as one of the first Preserve America communities.

2007 The James Madison Inn was built, becoming a landmark in the redevelopment of downtown Madison.

2008 Population of Madison: 4,405;
Population of Morgan County: 16,283;
Population of Georgia: 9,685,744 (estimates).

The *Morgan County Citizen* was recognized by the Georgia Press Association as best weekly of its size for the third time.

2009 Morgan County High School Principal Mark Wilson was named National Principal of the Year.

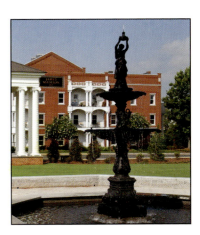

Madison Main Street celebrated its twenty-fifth anniversary as a National Main Street community.

October. Madison was selected to receive the highest award, "Excellence in Downtown Development," by the Georgia Downtown Association and the Georgia Department of Community Affairs, for excelling in all areas of measurement—organization, design, economic restructuring, and promotion—based on a historic preservation ethic and exemplifying a model to uphold.

December 12. The city dedicated its bicentennial legacy project, Madison Town Park, a public-private venture creating an outdoor event venue for downtown.

Madison Architecture and Preservation

For many years in Madison, architecture and historic preservation have gone hand in hand. Emblematic of this relationship are the museum houses of Madison, the earliest being the Reuben Rogers house, a two-story frame dwelling built about 1809 in a vernacular style called Plantation Plain. Other museums representing small houses from different eras of Madison's history are the Richter Cottage, Rose Cottage, and the African-American Museum. On South Main Street Heritage Hall is an excellent example of a white-columned Greek Revival house, and in the Madison-Morgan Cultural Center a recreation of the parlor at Boxwood is decorated with the authentic high-style furnishings from that notable home. Throughout the town are the ongoing public and private efforts to preserve, restore, protect, and revitalize historic architecture, a perennial endeavor, indeed.

A major step forward in those efforts began in 1974 when the Madison Historic District was placed on the National Register of Historic Places of the United States Department of the Interior. Then, in 1976, when the 1895 Romanesque Revival grade school on South Main Street was adapted to become the Madison-Morgan Cultural Center, the town had a major historic preservation symbol in its very midst. There could hardly be a better example of historic architecture and preservation than this.

In April 1988 the Madison Historic District Preservation Commission was established as part of the National Register Historic District program. At the same time the district was officially designated an ongoing project by the Madison mayor and city council. A ten-lecture series for district property owners and residents took place at the cultural center in the fall of 1989. In 1990 the commission published William Chapman's *Madison Historic Preservation Manual*. This excellent manual includes a copy of the Madison Historic Preservation Committee Ordinance and other technical aspects of the undertaking, including a list of properties in the Madison Historic District.

All these efforts were intended to help preserve Madison's long-standing historic character, as time and new generations bring inevitable changes and additions to the historic district. Protections are established that foster appropriate longtime maintenance and ongoing care to Madison's splendid treasury of historic architecture, gardens, and landscape features. This is not, of course, to freeze time, but to encourage appreciation for Madison's enduring qualities as a historic built environment.

As part of these preservation efforts, the Morgan County Foundation, Inc., which operates the cultural center, published a 140-page volume, *Madison, Georgia, an Architectural Guide*, in 1991. Each site is pictured in black and white, described architecturally, and assigned a number that shows on a schematic map. The *Guide* has "Notes on Architectural Styles in Madison," with a brief introduction that remarks: "While Madison may be better known for its Greek Revival buildings, a number of other styles deserve notice: The Gothic Revival, Italianate, Queen Anne, and Romanesque Revival styles, among others, play a lively role in the Madison streetscapes." Among the other styles mentioned are "Colonial Revival/Classical Revival (c. 1860–present), the Beaux-Arts and Neoclassical Revival, the Prairie House and the Bungalow."

The following portfolio, like the *Guide*, features major residential examples from all the district's areas and styles. Unlike the *Guide*, examples from Morgan County are included.

Madison and Morgan County are blessed with a treasury of architecture that its exemplary historic preservation standards are keeping alive, generation after generation, as page after page of this bicentennial book attests. After all, in more ways than one, the people of this beautiful Georgia town and county have proudly made this book possible.

The Baldwin-Williford-Ruffin house (c. 1850) on South Main Street was once the classroom building for the Madison Collegiate Institute.

Antebellum Architecture
Log Cabins to Neoclassical

The Madison Historic District is best known as an antebellum, Old South town, somewhat like Natchez, Mississippi, a place of pilgrimage to experience authentic history and southern charm, well-preserved and presented, but having evolved into the New South nineteenth, twentieth, and twenty-first centuries. Madison is known far and wide as a small-town urban fabric of architectural treasures, beautiful gardens, and tree-lined streets.

That reputation is well deserved. It is a consistent reputation from Madison's actual antebellum days, c.1809–c.1861, some half a century. After the war, in the postbellum era, this reputation continued and persists to this day. Georgia historian and state archivist Lucian Lamar Knight wrote in his *Georgia's Landmarks, Memorials and Legends* (1914): "Stately homes of the old regime are still standing in Madison, but while the past is revered, the town's progressive enterprise is typical of the New South."

One could teach a course in social, architectural, and landscape gardening history using Madison and Morgan County as a living laboratory. There are textbook examples of almost all the southern American styles from the antebellum era, and since Madison was founded as the young nation's taste turned toward the classical forms of ancient Rome and Greece, its architecture evolved along with a progression of styles as they worked their way from the great coastal cities out into the frontier towns, overlapping with economic surges as Georgian gave way to Federal, then Greek Revival, and on to Gothic Revival and Italianate. Madisonians of a conservative nature tended to apply one style over another as time passed, rather than tearing down to build anew, and many of the town's old homes bear testament to this practice.

The earliest houses built by white settlers in Madison were quite

> *The Greek Revival ... represents one of the highest cultural achievements of the state and one that still retains a powerful hold not only on the people of Georgia, but upon southerners generally. Developing concurrently with the more monumental Greek Revival, was the "plantation plain style" built ... with simple unpretentious detail eminently suited to the expanding frontier.*
>
> Frederick Doveton Nichols
> **The Early Architecture of Georgia**

Madison: A Classic Southern Town

Opposite: Carter-Newton Greek Revival house on Academy Street (1850). Above: An 1820 drawing of a town being carved from the Georgia forests. Below left: The Henry Lane dogtrot log cabin in Morgan County (c. 1810). Below center: Single-pen cottage from late nineteenth century, now a guest house on Dixie Avenue. Below right: Double-pen cottage (c. 1820) moved from South Main Street lot, now at Bonar Hall.

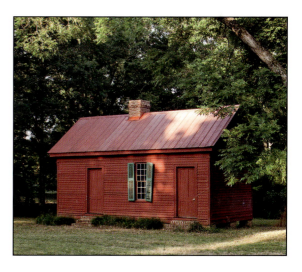

basic. Many were one- and two-room frame constructions no doubt similar in form to the ones built on Yamacraw Bluff in 1733 by men and women brought by James Edward Oglethorpe to establish Savannah. In Madison, as in similar frontier towns across piedmont Georgia, these simplest of houses were built alongside frame and log cabins with a "dogtrot" (two rooms separated by an open hall, or breezeway); two-room, central-hall cottages; and the slightly more substantial, two-story homes in the vernacular Plantation Plain style. Such house plans had practical value for those of meager means, so they continued to be used up through the nineteenth century, and examples from al-

Above left: The Reuben Rogers Plantation Plain style house (c. 1809–10) on East Jefferson Street. Above center: The Cooke house on Academy Street, an early nineteenth-century-center-hall house, was remodeled and expanded over time. Above right: The Saffold house on Second Street is apparently an early nineteenth-century Federal style house later modified with a Greek Revival addition. Below: Hilltop (c. 1838), a traditional center-hall plan with Greek Revival details, including the one-story Doric portico.

most every era can still be found in Madison.

Plantation Plain generally refers to an indigenous form particularly suited to the long, hot summers of the Deep South: to encourage ventilating breezes, such houses were only one room deep with windows on three sides, the two rooms usually separated by a central hall with doors at each end; to offer shade, one-story, shed-roofed porches sheltered the front elevation and often the rear; to provide space (given the previous considerations), there were two stories, and sometimes the rear porch would be enclosed. Chimneys were at the gable ends, and kitchens were detached or connected by breezeways to the main house.

The Plantation Plain was such a logical form for the region that in Madison it can be found trimmed with Post-Colonial stylistic details from the Federal period forward. By 1809 the Georgian style of the neoclassical era had given way to the Federal, so the earliest Madison examples with any ornamentation usually had Federal influences. Later such houses could have Greek Revival door surrounds and mantels and eventually Gothic Revival trellis porches or Italianate eave brackets.

Apparently the oldest Madison survivor of the Piedmont Plantation Plain style is the Reuben Rogers house, an 1809–10, furnished museum, on its original site across from the town square. Some unadorned Plantation Plain houses were later embellished with stylish trim or even subsumed into grand and up-to-date expansions, and examples of early houses lie completely hidden inside some of Madison's most familiar landmarks.

For example, the earliest part of the mansion known as Thurleston, on Dixie Avenue, is a Plain style house from 1818, to which was added a three-gabled neoclassical façade in 1848. The Stokes-McHenry house on Old Post Road holds within its squared-off plan and sophisticated details a basic double-pen house from the 1820s. Perhaps most surprising is the Foster-Roman house on Academy; onto the thin walls and small rooms of a very early house was attached an exuberant Neoclassical Revival remodeling.

As the cotton economy of antebellum Morgan County grew and planters and merchants

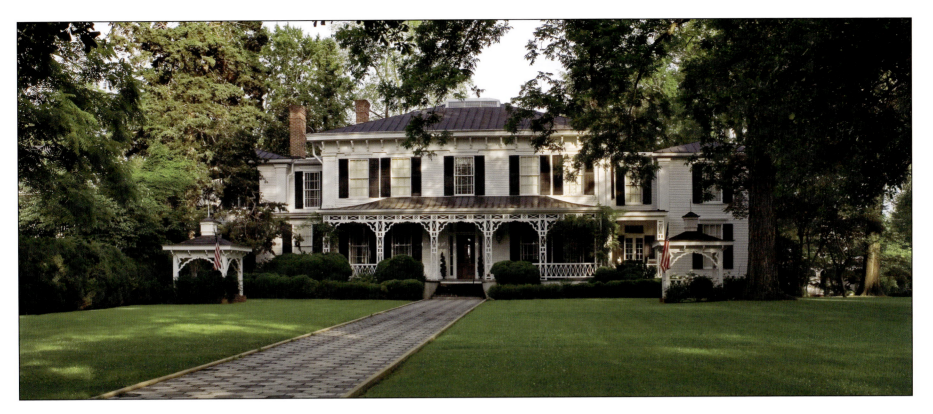

*Above: The Broughton-Sanders house (c. 1852) on Academy Street combines Greek Revival, Italianate, and Gothic Revival elements.
Below: The Old Post Road façade of Boxwood (1851–52).*

prospered, new houses went up in Madison in the famously popular Greek Revival style, and older house fronts were cosmetically augmented with an application of columned porticoes. Because of this economic boon and its resulting building boom, Madison's classical signature was set, and because of her citizens' proud commitment to preservation, it has remained so to this day.

A drive through town on Main Street passes a series of examples from King Cotton's reign beginning with two-story Honeymoon at the southern edge of the historic district and continuing with one- and two-story cottages and mansions with barely a pause before reaching the town center. Then on North Main the parade resumes with a concentration of Greek Revival designs remarkable in quality and number. Madison in 1850 was truly "in high cotton."

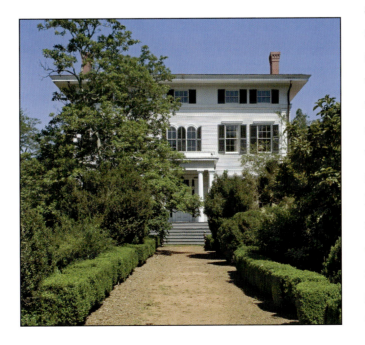

Madison's antebellum transition from Greek Revival neoclassicism to the picturesque Gothic Revival and Italianate is best illustrated in a three-house arrangement along Old Post Road and Academy Street. Boxwood, the neighboring Broughton-Sanders house, and the Stokes-McHenry house across the street are each in some way manifestations of the trend away from the somewhat severe formality of the Greek Revival to the thinner lines and romantic curves and arches of its stylistic descendants. Boxwood, with its single-story classical portico saluting Old Post Road and its fanciful trellised verandah facing Academy Street, represents the zenith of antebellum house design in Madison. Though built almost a full decade before war erupted, it is the last great residence on the morning side of cotton's economic mountain.

ANTEBELLUM ARCHITECTURE 67

Reuben Rogers House
179 East Jefferson Street
1809–10; remodeled 1873

This two-story frame house across from the town square and next to the Morgan County Courthouse became a City of Madison house museum in 1992. At that time, it had had numerous private owners since Reuben and Elizabeth Rogers built it in 1809–10, as Madison was founded.

In 1817 Thomas Norris, a Revolutionary War veteran, and his wife, Sarah, bought the property. Norris died the next year and was buried in the city cemetery, but Sarah lived here until she died in 1824. Sarah Norris added the two rear shed rooms in 1820. Ownership then included a Civil War veteran, Felix Martin. In 1868 Martin L. Richter bought the house, remodeling it in 1873. The front porch and some other details were as we know the house today. In 1886 Richter sold the house to John H. Hunter, whose family owned the place for over ninety years.

After the property was a dentist's office for several years, the Morgan County commissioners purchased it in 1992. An intergovernmen-

Opposite: The Rogers house sits on Jefferson Street between Rose Cottage and the Morgan County Courthouse. Above: Living room. Right: Parlor with faux-grained wainscoting.

tal agreement allowed the City of Madison to restore and furnish the house so it could be opened to the public under the management of the Morgan County Historical Society.

The historical society described the Rogers house as a "fine example of Piedmont Plain style architecture. . . . It looks as it probably appeared in 1873. The furnishings represent the mid-nineteenth century. The house predates the present Morgan County Courthouse by almost one hundred years."

EDMUND WALKER TOWN HOUSE
484 ACADEMY STREET
C. 1838; ALTERATIONS IN 1978 AND 1985

If proof for the popularity of this two-story antebellum house type (or style) is needed, this Madison house can be cited. It was popular when built and has remained livable and stylish today. This house, in fact, has a twin out in the county, and both houses have survived year after year, serving many generations.

Edmund Walker, whose family came to Morgan County in 1810, built this in-town house in 1838, a virtual duplicate of the family's 1815 plantation house on the Monticello Road. It was practical at the time for plantation owners to have a house in town to take advantage of educational and social opportunities while traveling as early commuters to their farms for business.

By 1838, when the Madison house was built, the Greek Revival style with its formal columns and pediments was sweeping the South, not to mention the rest of the expanding nation, but the Plantation Plain style, sometimes called the *I*-house, was holding its

Geographer Dr. Fred B. Kniffen, a specialist in folk housing, reports in *Readings in American Vernacular Architecture* (University of Georgia Press, 1986) that he coined the term *I*-house in 1936 in recognition of the two-story houses that early settlers from Indiana, Illinois, and Iowa built in rural Louisiana before the Civil War; the *I* seemed appropriate to him given the tall shallow form, which in Georgia has been called Plantation Plain. It has been a serviceable and simply elegant house style for over two hundred years.

Mr. and Mrs. Jerry Canupp bought and restored the town house in 1978. Then, in 1987, Dr. and Mrs. Donald Lane bought the property as an eventual retirement nest, moving here permanently in 1989. Dr. Lane commuted to his Atlanta medical practice until he retired in 2000. The house appealed to the Lanes as serious antique collectors;

own. the Madison *Architectural Guide* described this home in 1991 in this way:

> This two-story, one room deep dwelling with end roof gables [and end chimneys] illustrates a common early-nineteenth-century vernacular form known as the I-house. The added element of the rear shed creates a Southern variation now called Plantation Plain style. The balustrade, chamfered posts, and spindlework frieze of the front porch show a later Victorian influence. (page 117)

in fact, in Dr. Lane's retirement, he is involved with a Madison antiques store specializing in southern decorative arts.

The house the Walkers built out in the county was restored in the 1950s by Georgia decorative arts expert Henry D. Green. Now called Little River Farm, it has since been the rural retreat for several families who value the pace and quiet of Morgan County.

The Walker Town House sits on a large lot in Madison, where children ride bicycles on the sidewalks and horses in the lanes.

72　*Madison: A Classic Southern Town*

Opposite page: The Lanes have appointed the Walker Town House with a combination of family items, collections, and handsome American antiques. Top left: Living room. Bottom: The rear segment of the central hall was originally an open porch and now houses Dr. Lane's collection of antique toys. Top right: View from the entrance hall into the dining room.
Above: Dining room looking into the kitchen. Left: Garden.

Richter Cottage
c. 1830; enlarged and remodeled 1852
490 West Washington Street

The family of Charles William Richter (1807–84) came to Madison in the 1840s from the island of St. Croix in the West Indies. Richter, the son of a political exile from Germany, was a watchmaker and silversmith. It is thought that when he bought this property in 1852, it included a one-room structure used for an office or work room. Starting on a rise of ground above highway 83 on the road to Bostwick, the property extends to Tanyard Branch in the rear. The one-room shed likely was originally used in some way in relation to the tanyard operation on the creek.

Richter hired local builder James Morrison Broadfield to enlarge and modify the building. The result was a charming and practical, somewhat simplified Greek Revival cottage with a typical central hall plan.

In the 1970s, when the cottage was about to collapse, Dr. Josephine Hart Brandon, who resided nearby in the Atkinson family home, encouraged her seventh-grade students to make the early structure an American history project. The Morgan County Landmarks Society, chartered in 1976, the city of Madison, and many citizens contributed toward its restoration, which was largely completed in the mid-1990s.

The cottage and its grounds are the ongoing focus of a school heritage education program, kindergarten through twelfth grade and the furnished house museum is open to the public as an educational exhibit. The gardens around the property recreate an antebellum Madison landscape, especially the picket fence and early plant materials. A plain vernacular version of Greek Revival, the Richter Cottage house museum speaks eloquently of the life of early Madison and Morgan Country.

Opposite page, top: The front of the Richter Cottage looks out over West Washington Street (highway 83). The interiors of the cottage are furnished to complement the straightforward design. Opposite page, lower left: Bedroom. Lower right: Central hall. Above: View from central hall into dining room.

Stagecoach House
549 Old Post Road
c. 1810–20; remodeled c. 1845–6

Top: The Stagecoach House is thought to have been an inn on Old Post Road. Above: The wide central hall separates the two original first-floor rooms. Opposite page: Now a comfortable family room, this was a kitchen when the Eskews purchased the house.

In the *Madisonian* newspaper's 1992 edition of *The Madisonian, Architecture and History*, this is described as the Cornelius Vason house: "One of the oldest structures in Madison, this house was an inn when Old Post Road was a part of the stagecoach route between Charleston and New Orleans." In the 1996 *Teacher's Resource Guide* to Madison, Susan L. Hitchcock refers to the place as the Neil Vason house and writes, "This old inn [was] built by John Colbert." Hitchcock quotes old newspaper accounts of the place. One from the July 27, 1900, *Madisonian* told of its "lovely old-fashioned flower gardens." Another from the Albany *Herald*, October 23, 1939, read: "A large two-story frame house . . . sits near the sidewalk. The veranda has tall columns and a wide doorway leads into the hall which extends the length of the house. The large rooms with extra wide windows give a spacious antebellum atmosphere to the roomy old house."

The current owners of the property are Glenn Eskew and Pamela Hall. A native of Birmingham, Alabama, Dr. Eskew is a an associate professor of history at Georgia State University, commuting between Madison and Atlanta. In a detailed account of his and his wife's Madison home, he summarized what has been researched and written over the years about the place, some of it by him and some by Madison's historian/archivist Marshall "Woody" Williams. Eskew calls his home the Stagecoach House.

Historians have concluded that the earliest part of the house dates from 1810–20 and was probably moved to the site. Glenn Eskew writes, "The original Piedmont Plain style house was extended to a four over four with a vernacular Greek Revival porch attached, which probably took place in 1845/46." Dr. Eskew says that Neil Vason Sr. acquired the property in 1941 and that it passed to Vason's sons, Neil Vason Jr. and Wayne Vason, in 1975. From Wayne Vason it passed to Glenn and Pamela in September 2002, when Eskew and his wife, a practicing physician, began an extensive restoration/renovation, among other things putting on a new roof and creating a family room in the Vasons' old kitchen. Glenn and Pamela are raising their two sons here, as life proceeds apace for the old stagecoach inn.

Cedar Lane Farm
Sandy Creek Road
c. 1830; restored 1967

Henry Hilsabeck, a Moravian German living in North Carolina, in 1812 acquired several hundred acres in Morgan County, Georgia. By the 1830s he had built this Plantation Plain version of a Greek Revival house on Sandy Creek Road north of Madison. Hilsabeck was a well-to-do planter when he died in 1845, and his plantation lands were more than twice the size he started with.

The house is now the home of Jane Symmes, a member of the first board of the Georgia Trust for Historic Preservation and the widow of John C. Symmes, a graduate horticulturist, nurseryman, and naturalist. The Symmeses acquired the deteriorating house in 1966 and moved there from Atlanta, naming the place Cedar Lane Farm, which essentially comprised the same two hundred acres that Hilsabeck had started with in 1812.

When the house was added to the National Register of Historic Places in 1971, it was considered to be a model of historic preservation and restoration, inside and out. The Symmeses had also restored the grounds, featuring native plants and flora, and planted a fenced antebellum-style boxwood parterre garden to the side. The house was featured in *The Architecture of Georgia* in 1976 and *Landmark Homes of Georgia* in 1982. For many years Cedar Lane Farm was a commercial nursery where southern plants and trees were cultivated and propagated for sale throughout the region.

The text in *Landmark Homes* described the place as an "unusual example of the Greek Revival vernacular with Plantation Plain style

Above left: The entrance to the Symmeses' restoration of Henry Hilsabeck's house. Above right: The kitchen yard overlooks meadows and gardens. Opposite: A boxwood parterre garden on the west side of the house.

78   *Madison: A Classic Southern Town*

Above and opposite: The unpainted pine boards of the dining room and central stair hall suit the house and its piedmont landscape.

characteristics. The profile of the house is Plantation Plain but the plan and detailing is Greek Revival. Mantelpieces are the main cabinetwork, unless one counts wide, unpainted 'pine paneling.'" The house faces north, so the need for a front porch was minimized.

The way the house is furnished emphasizes the pre-1840 aspects of its architecture, with a c.1800 Sheraton sofa in the parlor (the only room in the house ever to be painted). Other furnishings are country Federal. Little or nothing is evident of the Empire and mid-Victorian era, which the furnishings of antebellum Madison sometimes bring to mind.

Cedar Lane Farm appeared in the spring 1981 issue of *Southern Accents*, which detailed the restoration procedure Jane and John Symmes followed to bring this antebellum home to its present award-winning condition. We expect Henry Hilsabeck would approve of what has happened to his restored Plain style homeplace.

Hilltop
543 North Main Street
c. 1838

Situated on one of the highest sites in Madison, on a street with a number of other handsome antebellum homes, Hilltop was apparently built about 1838 by Thomas J. Burney for his new bride after his first wife's death. It is pictured in *The Early Architecture of Georgia* (1957) as "Hilltop, the Burney-Lambert House." In 1923 it became the home of Mr. and Mrs. Ezekiel Roy Lambert Sr.; in 1925, son Roy Jr. was born here. Both Lamberts, father and son, practiced law. The younger Lambert opened his office in Madison in 1950, and four years later he married Christine Davis, who grew up in Atlanta.

Following his mother's death, Roy and Chris renovated and restored his family home, and they moved in on October 1, 1966. Roy served first in the Georgia Senate and then the Georgia House of Representatives for twenty-two years, until 1984. He also served as president of the Georgia Trust for Historic Preservation for several years in the late 1980s.

William Chapman in his *Madison Historic Preservation Manual* captions a photograph of Hilltop as a "Greek Revival portico on an otherwise 'Plantation Plain' I-house." Chapman says, "The portico may or may not be original to this greatly significant 1830s property" (page 30). The house itself suggests, however, that it may have been built in

an L shape, rather than the standard I of the more Plain style plan, and the date of construction would certainly allow for the Greek Revival details—the portico and the transom and sidelights, for example—to have been original.

When Roy Lambert died in 2008 at age 82, his obituary said that "he lived in the house on North Main Street in Madison where he was born. . . . his life was centered in Madison and Morgan County." Chris Lambert, who was the first chair of Madison's Historic Preservation Commission, still makes Hilltop her home, which she often shares with the community as she continues the Lambert tradition of civic leadership.

Opposite: Hilltop stands on a rise among other classical homes on North Main Street, where large lots and a canopy of mature trees give a rural feeling only a few blocks from downtown Madison. Above: Entrance/stair hall. The interiors at Hilltop display refined decorative details within a modified Plain style plan.

Left and top: The Lambert living room is appointed with family pieces and art collected over many years. Above: Desk and painting. Following pages: Dining room with view into breakfast room.

HILLTOP 85

Thurleston
847 Dixie Avenue
1818; addition/renovation in 1848

This impressively sited, large-scale Madison manor has traditionally been called Thurleston, after Thirlestane Castle in Scotland, about which Sir Walter Scott wrote a poem. Located on a thirty-five-acre tract, Madison's Thurleston evolved from a smaller Plantation Plain style frame house, built in 1818 and moved to this site in 1841 by the John Walker family from Three River Farm, their plantation out in the county.

The front part of Thurleston was added in 1848 by Dr. Elijah E. Jones, a pioneer Morgan County citizen and a large landowner. His architect is said to have been one Benjamin Peoples, about whom little else is known. The original house was more than doubled in size, giving the entrance façade, with its three matching gables and giant pilasters, a monumental and at the same time eclectic appearance, mixing Georgian and neoclassical Greek Revival with a touch of Gothic Revival because of the three gables.

It has been the home of Mr. and Mrs. Clarence Whiteside since 1983. The Whitesides moved to Madison from Florida, says Kathy Whiteside, to raise their three children in an idyllic smaller community "with a town square and where you know neighbors by their first names." Madison more than qualified once they saw Thurleston and its setting; they had first come here as part of one of Clarence's hunting trips to Georgia. Beginning in 1982, the Whitesides restored the property and made appropriate additions to the house under the guidance of one of Georgia's most accomplished architects, classicist Norman D. Askins of Atlanta. Mrs. Whiteside rebuilt and expanded the Thurleston landscape, as she says, "on the bones of the original

gardens designed by Eatonton horticulturist D. Benjamin Hunt, who was a family friend of the Butler sisters."

The Whitesides bought the property from Colonel Harold Wallace, who had inherited it from his mother. Thurleston had come to her from Virginia Butler Nicholson, a granddaughter of Madisonian Senator Joshua Hill and a niece of Miss Bessie Butler, one of the longtime owners and residents. A letter dated January 9, 1925, to Bessie

Opposite page: Thurleston from Dixie Avenue is viewed down a long allée of boxwoods. Above: Side views of Thurleston reveal some of the construction history of the house. The north side (above right) clearly shows the profile of the earlier Plain style house hidden within the fabric of the later additions.

Opposite page: The central hall of Thurleston looks out on the boxwood allée.
Above: The stairs rise from the south end of the transverse hall.

Butler from a prominent Atlanta socialite, Mrs. Hugh (Josephine Inman) Richardson, after a visit to Madison, tells the story of Thurleston to this day: "The hours spent in your wonderful old home were indeed happy ones. . . . May we bring the children down to see your garden when it is in bloom? I want them to have a glimpse of the days of long ago which your home so perfectly exemplifies" (*Teacher's Resource Guide* [1996], page 8).

Opposite and above: Greek Revival details adorn the parlors and central hall of the 1844 addition to the original Plain style house. Right: The Thurleston gardens are many and varied.

BARNETT-STOKES HOUSE
752 DIXIE AVENUE
c. 1830

Located just down the road from Bonar Hall and Thurleston, this antebellum Greek Revival raised cottage is a form more typical of the southern coast and is unusual in Madison. It is also similar to a type described by Frederick Doveton Nichols in *The Early Architecture of Georgia*: "The Sand Hills area of Augusta was originally used for summer homes, and a style of such definite, architectural character evolved there that the type became known as the 'sand hills cottage.' It consisted of a story-and-a-half house with dormers and a center-hall plan, usually raised on a high basement to keep the house cool and dry." In Perkerson's *White Columns in Georgia* is the notice, "There is also the charming raised cottage of Mrs. Mary Barnett Stokes and Miss Katie Porter Barnett." Louis McHenry Hicky, in her *Rambles through Morgan County*, speaks of it as the home of Ann Fannin Stokes (daughter of Mary Barnett

Opposite top: The deep galleries of the Barnett-Stokes house offer afternoon shade and outdoor living space. Opposite bottom: The north elevation with geometric gardens. Above: The rear of the house looks out over a rolling terrain and mature plantings.

Stokes) and "almost a replica of Beauvoir, the Biloxi, Mississippi, home of Jeff Davis."

The house is rich in local history and lore. Famed composer Stephen Foster is said to have been a guest there, and a slit in the front door is supposed to have been made by an angry, sword-wielding Yankee officer. The Madison (Confederate) jasmine cultivar was discovered on the grounds. Named and propagated by the Symmeses at Cedar Lane Farm, it is now grown throughout the piedmont.

For many years, this was the home of Miss Ann Fannin Stokes (1892–1983), the last member of the Barnett-Stokes family to live here.

A subsequent owner, Dottie Billingsley, undertook the extensive renovations of the house and gardens in the late 1980s.

In 1994 Shandon and David Land and their younger son, Jonathan, moved from Florida, attracted by the historical character of the community and this house in particular. The couple made appropriate interior changes, furnishing their home with antiques, including a fine collection of original McKenney and Hall Native American portrait lithographs and pieces reflecting their Florida roots and David's work abroad. Shandon serves on the Historic Preservation Commission. David is chairman of the Historic Madison-Morgan Foundation.

The interiors of the Lands' house blend modern living and comfort into the architectural context of their historic home. Above: The dining room and central hall. Opposite page, top: The family room and rear porch. Bottom left: Downstairs kitchen. Bottom right: In the lower stair hall hang McKenney-Hall portraits on wide pine paneling above a painted and stencilled floor.

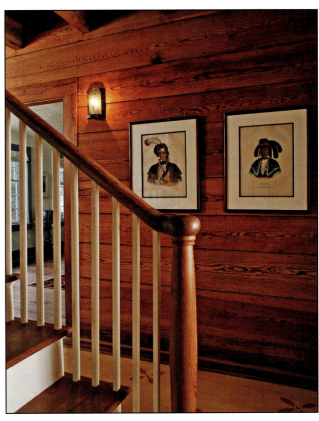

ROBSON-MASON HOUSE
1041 DIXIE AVENUE
C. 1830; REBUILT IN 1940S

The home of Mr. and Mrs. Charles L. Mason Jr. was originally a farmhouse version of Greek Revival style built by John Robson in the 1830s. Robson is known to have built other houses in the area but lived in this house out on Dixie Avenue until sometime in the 1840s. The railroad running along the western edge of the property also marks part of the current town limits, and this is just one of several places in Madison proper where one can see horses cantering in pastures and cattle grazing alongside classical homes.

In the late 1930s the house was damaged by a fire that partially destroyed the upper floor, which was not rebuilt. The father of Charles Mason Jr. bought the property in 1941 and restored and renovated it. Ten years later he raised it on a high basement, and the place today looks as if it were always a one-story Greek Revival cottage, with a low-pitched hipped roof.

Charlie and Janet Clay Mason have continued to make improvements for twentieth-century comfort. Janet is an artist, musician, and civic leader and a member of the Historic Madison-Morgan Foundation. Charles continued the Mason family agricultural interests begun in Morgan County by his grandfather, Charles Ross Mason (1881–1975). Among the art works in the Masons' collection are pastels of their daughters, done by local Madison artist Martha Lower, whose portraits grace many of the homes in this portfolio.

Opposite: Entrance elevation. Above: The parlor is decorated with period pieces and family artifacts.
Below left: A swing set is festooned with wisteria blossoms in the spring. Below right: An evergreen garden of boxwoods and camellias.

Appreciation of art, light, and texture are evident in the Masons' home. Right: Dining room and entrance hall. Top: Entrance hall. Above: The den blends modern art with antiques and mementos.

Bonar Hall
962 Dixie Avenue
1839–40; Remodeled c. 1885

The present owner of this landmark hall, Alexander Durham Newton, is documenting its history from A to Z in his "History of Bonar Hall," which he continues to update as he makes new discoveries. The house and ten acres of the property were listed in the National Register of Historic Places in 1971–72 when this author was director of the Georgia Historic Sites Survey and worked on the nomination with Miss Therese Newton, the current owner's aunt, from whom he inherited the place in 1994. Miss Therese had herself inherited it, with her brother, Edward Taylor Newton II, in 1920. When Miss Newton died in 1994, she had lived at Bonar Hall for 74 of her 94 years.

One fact that Alex unearthed that was not mentioned in the Bonar Hall National Register nomination is the origin of the name Bonar. Josie Newton Bacon, Miss Therese's mother, had named the house for her great-great-great grandfather Charles Bonar, of Virginia. The oldest memorabilia in the house connects to the Virginia Bonars; a Bonar portrait hangs in the sitting room.

The rudimentary history of this exceptionally historic place is that it was built by John Byne Walker (1805–83) and his wife Eliza Fannin Walker, whose father owned the land. Walker was the successful and confident son of early Morgan County pioneers and an antebellum plantation millionaire, but Eliza's inherited wealth dwarfed his, and her personality was his match. According to John Walker's farm journal, the first bricks were "laid on February 25, 1839, starting with the brick kitchen, then the main house." They moved in fifteen months later. The house, sitting two hundred feet from the road, was, as Alex Newton described it, a "four-over-four traditional Georgian manor house with rooms 20' X 20' and eight fireplaces. The ceilings are almost thirteen feet on both floors, making the house feel much larger inside than it seems from the outside." Originally, there was a one-story classical portico, instead of today's two-tiered Victorian veranda, which was "added in the Broughton era." William A. Broughton and his wife Mary "Molly" Pou (whose family then owned Boxwood) made all the Victorian stylistic changes after 1880, when they purchased the house.

The formal gardens planted by the John Byne Walkers around 1850 still survive. The grounds are featured and described in the *Garden History of Georgia, 1733–1933*, as a "hundred acre tract of lawn, garden, and orchard with a summer house and orangery of matching design flanking the house." Another account mentions "seven varieties of magnolia trees and six varieties of boxwood."

Alex and his wife, Betsy Wagenhauser, originally of Dallas, Texas, in 1993 began the painstaking renovation/restoration, inside and out, of their Newton family landmark, an effort that still proceeds in 2009.

The expansive grounds of Bonar Hall include the "manor house," detached brick kitchen, flanking summer house and orangery, and several other dependencies.

BONAR HALL

Heritage Hall
277 South Main Street
c. 1811; extensive remodeling c. 1830s–40s

It is entirely appropriate that one of the finest of the Greek Revival houses on Madison's Main Street has been preserved as a house museum and historical society headquarters, open to the public and available for special events.

Integral to the Madison Historic District, the house originally stood just two hundred feet south, facing Central Avenue. Recent research indicates Dr. William Johnston (1784–1856), a wealthy land and slave owner in Morgan and adjacent counties, may have built the house as early as 1811 in a much simpler form. He sold the property in 1830 to Dr. Elijah Evans Jones (1795–1876), who apparently began enlarging the house and remodeling it in the Greek Revival style, attaching a monumental portico and adding neoclassical detailing to some of the woodwork. The front entrance duplicates an illustration from an influential Greek Revival design book, *The Young Builders General Instruction*, by Minard Lafever (1798–1854), published in 1829, which helps pin down the date of this high-style Greek Revival renovation. The four fluted Greek Doric columns with two square fluted corner columns, *in antis*, are a unique survivor from Madison's antebellum era.

Steve Turnell purchased the place in the latter part of the nineteenth century and sometime after 1909 moved the house and rotated

Opposite page: The Greek Revival façade was likely added in the 1830s or '40s when the house was remodeled. Above: The dining room and parlor are connected by tall pocket doors in a neoclassical frame.

it to face Main Street. His family sold the original lot on Central Avenue for the construction of the First United Methodist Church, which was completed in 1915. In 1923 Turnell opened the house as the Traveler's Inn, which closed in 1933 following a fire.

Mrs. W. Fletcher Manley Sr. acquired the house in 1946. After she died in 1977 at age 93, this Madison landmark was given to the Morgan County Historical Society in her honor by her granddaughter, Mrs. George (Susan Manley) Law Jr.

Left: High ceilings and a wide, ventilating central hall helped keep homes cool in summer. The hall also served as additional social and living space. Above: View from the hall into the parlor framed by the strong lines of a Greek Revival door.

MARTIN-BALDWIN-WEAVER HOUSE
488 NORTH MAIN STREET
1850

The Parthenon, completed in 432 B.C., was architectural perfection in the classical period of Greece. Antebellum Greek Revival in Madison first appeared in the 1830s, then reached full expression in the 1840s and '50s with the construction of a number of temple-form houses about town. Several of these can be seen on a drive up the main street through Madison, from Honeymoon on the Eatonton Road and Heritage Hall near the town center to this house on North Main. Medora Perkerson in her *White Columns in Georgia* (1952) says, "There is the Thomas Baldwin house, originally the Felix Martin house with its fine Doric columns" (page 55). William Chapman in his Madison *Manual* describes it as a "'classic' Greek Revival house, built in 1850 by Felix Martin. It has Doric order columns, a cantilevered balcony, and a typical Greek Revival entry, with transom and sidelights" (page 7). Chapman also says that it is "probably the most significant 'pure' example of a Greek Revival temple front house in Georgia" (page 59). Also significant is the fact that only three families have owned this outstanding neoclassical design in its 159-year history.

Felix Bryan Martin, of French Huguenot ancestry, and his wife, Margaret Ann Fears, built the house about 1850 for their daughter, and upon her husband's death it was sold to Judge H. Walter Baldwin, who was active in the Confederate government as an assistant to Alexander H. Stephens. Baldwin's son inherited the house, and it remained in his family until the 1950s when, in time, it became the home of Russell and Dr. Rose Ann Weaver, who restored it, adding a classical porte cochere on the north side. The interior Greek Revival woodwork, including the mantelpieces, is an especially handsome original feature of this landmark's antebellum architectural style.

The façade of the Martin-Baldwin-Weaver house is perhaps the finest example of a temple-form Greek Revival design in Madison.

*Left: The parlor is serene, with strong details and tall windows.
Top: The dining room and parlor have similar proportions separated by the central stair hall (above).*

Honeymoon
928 Eatonton Road
1851

Built in 1851 by Charles Mallory Irwin, a well-known Baptist minister and political leader, this house was named Honeymoon by Mrs. Peter Walton Godfrey in the 1930s, after a family home in Florida. The publication of *Gone with the Wind* in 1936 and the white-columned relationship between Greek Revival Honeymoon and Scarlett O'Hara's Tara inspired a week-long house party with friends of Mrs. Godfrey's granddaughter, Caroline Candler of Atlanta (later Mrs. Lowry Hunt Sr.).

Appropriately included in *White Columns in Georgia* (1952), the house was described as "the Doric-columned home of Mr. and Mrs. Charles Candler . . . furnished with beautiful heirlooms" (page 54). Louise McHenry Hicky in her *Rambles through Morgan County* commented that Mr. Candler was a son of the illustrious Methodist Bishop Warren A. Candler (1857–1941).

Today, the house still contains Candler family heirlooms that bring back memories of the founding of Emory University, an outgrowth of the old Methodist college at nearby Oxford, Georgia. The college's

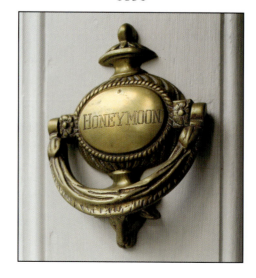

112   *Madison: A Classic Southern Town*

The grounds at Honeymoon have period outbuildings amid a landscape of fruit- and nut-bearing trees and flowering gardens with roses, daylilies, and irises.

move to Atlanta was led by Bishop Candler and his brother, Asa Griggs Candler, founder of the Coca-Cola Company. Honeymoon is currently owned by Lowry (Whitey) Hunt, Charles Candler Hunt, and their sister Suellen Sheppard, grandchildren of the Charles Candlers.

Madison's Honeymoon, Heritage Hall, and the Martin-Baldwin-Weaver house are three similar expressions of the symmetrical, two-story Greek Revival houses built in Georgia's piedmont towns from the 1830s to 1850s, with full façade porches and central halls lighted by entranceways with sidelights and transoms. These houses and many more throughout the region inspired Medora Perkerson to title her famous book *White Columns in Georgia*.

The Greek Revival interior trim is boldly proportioned and in deep relief. Portraits and mementos at Honeymoon are reminders of family service and traditions Above: Parlor. Opposite: Entrance/stair hall.

Baldwin-Williford-Ruffin House
472 South Main Street
c. 1850; remodeling in 1870s, 1970s, and 2003–4

This neoclassical Greek Revival landmark on South Main Street, adjacent to the Madison-Morgan Cultural Center, is the only building remaining from the first of Madison's schools developed expressly to enable female students to pursue higher education. A Baptist initiative, the school was chartered in 1850 as the Madison Collegiate Institute and served, as most similar institutions of the time, as a finishing school for young ladies. In 1851, perhaps to project a less provincial image, it was renamed the Georgia Female College.

This house, originally designed as the classroom building, included several large rooms for receptions and assembly. Soon it became the college president's home when a new brick classroom building was constructed immediately to the south. Both buildings are illustrated in a well-known lithograph of the college from about 1855. Classes were suspended during the Civil War, and the school never completely recovered. Then, after the classroom building burned around 1882, the college closed, and this house became a private residence.

Now the home of Robert E. Lanier, it has seen a series of owners over time. Thomas B. Baldwin Jr. owned it at the end of the nineteenth century before it descended to the Willifords. In 1967 it belonged to Mr. and Mrs. Durrell Ruffin. Robert Lanier, who previously had restored his great-grandfather's c. 1848 plantation home in Hancock County, spent two years restoring this splendid artifact from Madison's halcyon days before the Civil War.

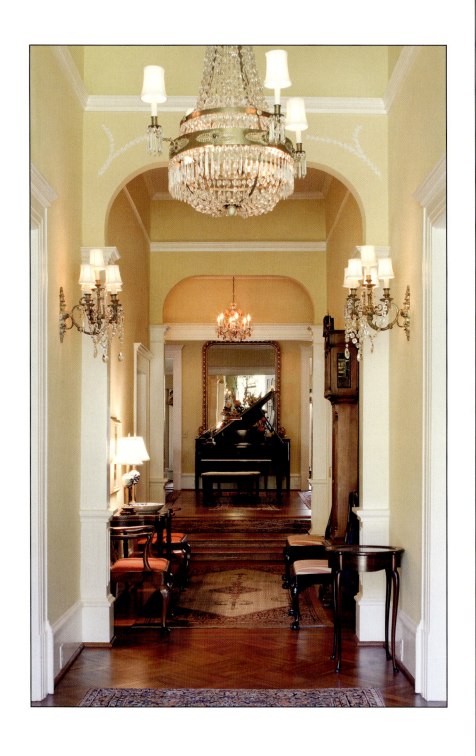

The house is so pleasantly proportioned on the outside that the interiors seem surprisingly spacious. Far left: The entrance hall leads to a large room the Laniers use for dining and entertaining. Left and below: Twin parlors flank the central hall. .

William Chapman's Madison *Manual* (1990) described this house as "Madison's most prominent example of the Early Classical Revival" (page 58). Others have called it Greek Revival. Nevertheless, it stands out on its prominent Main Street site, with its almost Palladian portico supported by four fluted Greek Doric columns, as a classical antebellum architectural landmark. It is somewhat reminiscent of Montpelier, President James Madison's own pedimented neoclassical home in Virginia.

The interiors of the Lanier home are formal yet inviting. Left and above: The living/dining room is several steps above the central hall and looks out onto a sunroom at the rear of the house. Below: Guest bedroom.

Billups-Tuell House
651 North Main Street
c. 1853; additions in 1893 and 2007

This graceful North Main Street house is a variation on the classical revival raised cottage, with its locally one-of-a-kind pediment and fanlight in the tympanum of a Doric portico. The wide entrance doorway, with transom and sidelights, and large nine-over-nine windows light a wide central hallway. The house was built about 1853 by General Jeptha Vining Harris for his daughter Susan, Mrs. Joel Abbot Billups. In the 1960s it was the home of Dr. and Mrs. W. R. Van Buskirk. The minister of the Madison Presbyterian Church, Dr. Van Buskirk established an apple orchard in the back of this property long known for its garden landscape.

In 2009 it is the home of Grady and Sally Tuell, who in 2006 began careful and extensive renovations to their house in keeping with the needs of a new generation. Grady's parents, the Hugh Tuells, bought this house on North Main in 1964. Today, in a pattern familiar to many Madisonians, Grady and Sally Tuell are raising their two young children in the house where he was raised.

An article about this property on February 1, 1907, in the *Madisonian* described an extensive garden here with blossoms (presumably camellias) "blooming beautifully since November," which confirms again what numerous writers have said about Madison gardens through the years. In the Georgia Department of Agriculture's 1901 publication, *Georgia Historical and Industrial*, O. B. Stevens wrote, "The ladies of Madison are noted for the taste displayed in the cultivation of the flower gardens which adorn so many of their charming homes" (page 770).

This early twentieth-century description of "charming homes" might be written about the entire antebellum stretch of Madison's Main Street, before it heads north toward the Classic City of Athens.

Opposite: The Billups-Tuell house façade is distinctive in form and detail. Above: The dining room mantel was replaced when the house was altered in 1893. When the Tuells' home in Mississippi was washed away by Hurricane Katrina, their china was the only thing they found.

Above: The living room mantel was also among the changes made when the house was remodeled at the end of the nineteenth century. The Greek Revival woodwork became a graphic element in the interior design when the old paint was stripped and a natural finish applied.
Opposite: Entry and central hall.

BILLUPS-TUELL HOUSE

Massey-Tipton-Bracewell House
611 North Main Street
c. 1854

Little is known about Nathan Massey, who built this house about 1854 on North Main Street. In 1957 Frederick Nichols, in his *Early Architecture of Georgia*, noted: "In Madison is the Tipton House, an example of a one-story house with a full Doric portico across the front and a pyramidal roof" (page 132). Fortunately, the house has not been altered over time, and the photograph accompanying that text, taken by Frances Benjamin Johnston in the 1930s, looks as if it could have been made today. Not long after the house was built, it was purchased by General Jeptha Vining Harris, who in 1853 built the handsome Greek Revival Billups-Tuell house next door for his daughter, Mrs. Joel Abbott (Susan) Billups. Since the 1980s it has been the home of Michael and Ruth Barrow Bracewell. Mike has served his community for years as probate judge. Ruth is devoted to the Madison and Morgan County arts community, especially the Madison-Morgan Cultural Center, and is a member of the Historic Madison-Morgan Foundation.

William Chapman writes in his Madison *Manual* that this is "an enormously significant Greek Revival

cottage" (page 60). There is a central entranceway framed by pilasters and the characteristic Greek Revival transom and sidelights. The columned porch wing on the south side of the main house blends perfectly with the front portico. This area of North Main Street is home to a number of significant antebellum classical revival houses built or improved during a brief but exceptional period in cotton's economic heyday.

Opposite page: Not all Greek Revival designs fit the stereotype of a grand, columned mansion. The Doric portico of the Bracewells' home is especially well designed and proportioned. The interior woodwork is crafted to fit the scale of the interiors. Opposite, bottom: View from the central hall into the living room. Above: Living room. Left: The central hall toward the front door.

Stokes-McHenry House
458 Old Post Road
c. 1822; enlarged and remodeled in 1840 and c. 1850

The delicate trellis porch of the Stokes-McHenry house is just one of the charming architectural elements of this eclectic composition. Below: Across Old Post Road from the Stokes-McHenry porch is the Broughton-Sanders house.

As the Greek Revival style was flourishing in antebellum Madison, it was bracketed in time and fashion by other national styles—the neoclassical Federal, Italianate, and Gothic Revival. The Stokes-McHenry house synthesizes elements of all these national antebellum styles in an eclectic and romantic composition developed as the house evolved over time. The Broughton-Sanders house across Old Post Road also shows influences of this nineteenth-century architectural exoticism led by such national figures as the tastemakers Alexander Jackson Davis (1803–92) and Andrew Jackson Downing (1815–52).

In 1952 Medora Perkerson in *White Columns in Georgia* described

The wide stair hall is set back from a transverse entry hall. The stair design implies a late Federal influence.

in words and photographs this house built by Judge William Sanders Stokes in 1822 and then occupied by succeeding generations of the same family. Perkerson's description was written in 1952, but the tradition continued. Mrs. Daniel (Louise Marion McHenry) Hicky, who was part of those numerous generations, said in 1971 in her book *Rambles through Morgan County*, "Seven generations of the same family have occupied this house, an early member having obtained the lot by lottery when the town was established."

The 1822 house is thought to have been one room deep and two wide, essentially the northern side of the current house; then the house was enlarged and modified in 1840 and again around 1850. As a result, there are hints of Federal, a strong influence of Greek Revival, and a light and distinctive trellis portico added when Italianate and Gothic Revival were becoming fashionable, as this house and its

The house is a veritable museum of almost two centuries of family history. Above and below: The small room on the north side of the entry hall is part of the original 1822 house. Right: Parlor.

130   MADISON: A CLASSIC SOUTHERN TOWN

Opposite and above: The remarkable second-floor hall with its oval concave ceiling is unique in Madison and perhaps in the state.

neighbors, Boxwood and the Broughton-Sanders house, attest.

Architectural historian Frederick Nichols admired the Greek Revival entrance in 1956: "A handsome wooden example of the period, inspired by Minard Lafever's vigorous and original variations on the Erechtheum doorway, is at the Stokes-McHenry house at Madison."

The 1992 edition of *The Madisonian, Architecture and History*, offers this description: "Built about 1822, the house is noted for its old furnishings, manuscripts, and first editions." At that time, in 1992, it was the home of Colonel Dan McHenry Hicky, the son of Louise Marion McHenry Hicky (1891–1984). Colonel Hicky inherited his mother's literary ability and is well known for his poetry. He and his children continue to keep the family traditions going in a house whose eclectic architecture seems to embody the romantic antebellum spirit of Madison.

Nathan Bennett House
1170 Dixie Highway
1850

This house was built in 1850 at a time when the construction of fine Greek Revival homes was booming in Madison. In addition to its forthright massing and picturesque details, it has an interesting history, and it retains a sense of those long-ago times.

The original owner, Milton B. Davis, lived here for only three years before selling it to John Cardwell, who left it to his daughter, Amanda. Her husband, Junius P. Smith, in 1863 lost the house to James Mann in a poker game. Mann tried to take possession of the house, and Amanda Smith naturally objected, but at the time in Georgia a husband's rights trumped his wife's even if she brought the property into the marriage. In 1869 the laws were changed so a woman's husband would no longer be able to act as her agent in such matters, but it came too late for Amanda Cardwell Smith. The case about her plight, however, is said to have influenced the passage of laws defining women's property rights more equitably in Georgia, which is one reason the house was put on the National Register of Historic Places in 1974.

The other reason for its registration was its architecture, which has changed little since it was built. To quote the nomination: "The main body of the residence has classical features in an Italianate mode. The one-story porch, spanning the width of the house, is an expression of Victorian fancy." This house represents the transition in antebellum Madison from Greek Revival to the Italianate eclecticism of the early Victorian period in American architecture.

When this house and five acres on the southern edge of town were nominated to the National Register, it belonged to Nathan R. Bennett. It is now the home of his son and daughter-in-law, Ray and Genia Bennett. Ray is a modern cattleman, for a time breeding and raising registered Holsteins and now brokering cattle. Genia was a math teacher for twenty-five years and four times was named Star Teacher by her Morgan County High School Star Students. The Bennetts' home has a familiar old-time comfort, where gentle breezes flow through the wide central hall and trains still rumble past in the night.

The Bennett house combines Greek Revival and Italianate elements.

NATHAN BENNETT HOUSE 137

Preceding pages: The Bennetts' entrance/stair hall is the most important room in the house. A conduit for light, air, sound, and traffic, it is the center around which the rest of the home revolves. All of the principal rooms open onto the hall, as it runs from the front porch to the rear, offering a breeze through the wide doors, which remain open when weather invites.

Above: Dining room.

Large windows provide every room in the house with ample ambient light and views of the grounds. The bedrooms are spacious, with high ceilings and proper proportions.

Boxwood
357 Academy Street
1851–52

"Built by the gentleman-farmer Wilds Kolb (1804–61), it is based on 'A Suburban Cottage in the Italian Style' in *The Architecture of Country Houses* by Andrew Jackson Downing, 1850. The trellis-work veranda almost duplicates the Downing design as does the overall form of the house." This was the way this three-story town house was described in 1982 in *Landmark Homes of Georgia*. Boxwood was new in 1851–52, and it continues to be preserved in 2009. In fact, the same family has owned the place since 1906, when it came into the possession and responsibility of the Newton family.

In 1982, when *Landmark Homes* was published, Boxwood was the home of Miss Kittie Newton, who was one of Madison's grand ladies of good will and something of a one-person chamber of commerce. She was mentioned fondly in *A Guide to Early American Homes* (1956): "Miss Kitty [sic] Newton, who lives in 'Boxwood,' one of the loveliest [Madison homes], has done us the great favor to say that she will try to make arrangements for interested visitors to see some of the homes listed here—at the owners' convenience, of course. . . . Miss Newton is also chairman of a tour sponsored by La Flora Garden Club during the latter part of April (not every year)." The article even gave Miss Kittie's telephone number and continued: "The house has elaborate and extremely elegant period parlors, whose every feature—carpets, draperies, chandeliers—is the unspoiled original furnishings of the days when Victorian could be regal. The house is a magnificent period piece, something to behold" (page 104).

Today Boxwood belongs to Mr. and Mrs. Floyd Newton Jr., who acquired it after the death of his Aunt Kittie in 1986. The parlor furnishings—twelve pieces bought by Wilds Kolb in 1856 and still in the original silk brocatelle upholstery—were donated by the Newtons to the Madison-Morgan Cultural Center, where they have been on display since 1998 for all to enjoy.

Astride the transition from the classical revival of the early and mid-1800s to the picturesque eclecticism of the latter part of the century, Boxwood's formal entrance faces Old Post Road with a classical one-story Doric portico, while its Italianate rear veranda looks out on Academy Street and the tall spire of the Church of the Advent. Fittingly, the parterres on the classical Old Post side of the house are laid

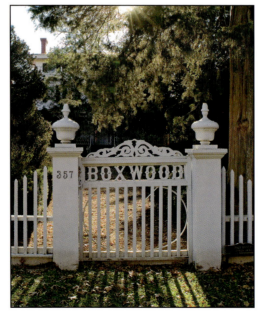

Boxwood is but one block off of Main Street, but it is firmly in its own world, a time apart from modern style or pace.
Above: Scenes at Boxwood. Opposite: The Old Post Road elevation.

out in straight lines and right angles, while those facing the lacy Italianate porch feature ovals and curves. Taking up half a city block, Boxwood derives its name from these elegant boxwood parterres maintained lovingly by each succeeding generation of owners.

Time marches on, but so do Boxwood and Floyd Newton Jr., who is in his nineties during the two-hundredth anniversary of Madison. Boxwood is indeed a landmark home of Georgia. In 2009, the Georgia Trust for Historic Preservation honored the Newton family's continuing enlightened ownership of Boxwood with an Excellence in Stewardship award.

Opposite: The Academy Street elevation with trellis porch and boxwood parterres in curvilinear patterns.
Above: Central hall looking out onto the Academy Street garden. The etched glass in the sidelights and transom is original.
The offset stair hall is unusual in Madison.

The parlor furnishings from Boxwood were donated to the Madison-Morgan Cultural Center in 1998. Left: View from central hall through trellis porch. Right: Double parlors with pocket doors. Above: West parlor looking out onto trellis porch. Opposite: East parlor.

Top, left: Second-floor guest bedroom. Above, left: Dining room. The mantel was rescued from a house being razed downtown. Above, right: The former servants' quarters are now a guest house. Opposite: The geometric shapes of boxwood parterres facing Old Post Road.

Postbellum Architecture
Folk Victorian to Neoclassical Revival

*Above: The Owen-Landry house (c. 1858) on Academy Street was renovated with elements of Italianate and Folk Victorian.
Below: Three Gothic Revival gables were added to the antebellum Atkinson house on Wellington Street in the late 1860s.*

From the Italian Renaissance days of the great architect Andrea Palladio (1508–80), domestic architecture was based on historical precedents and illustrated books. This is the way architectural ideas and styles spread throughout the world. Certainly it was no less true for the United States, both North and South, and for Madison, which was one of the most prosperous and cultivated of the antebellum Georgia piedmont towns.

After the Civil War to the South came Reconstruction and to the North, the Gilded Age, when the Vanderbilts and other plutocrats built palaces on Fifth Avenue in New York City. In general, the era is called Victorian, honoring England's Queen Victoria, who reigned from 1837 to 1901, some sixty-four years. Her era in architecture is often divided stylistically into early, mid, high, and late.

Architectural historians have had several terms for the styles that prevailed in the Victorian years. One that works well is picturesque

eclectic. The historical precedents that architects and builders used were indeed picturesque (the Vanderbilts, for example, liked the look of a French chateau) and the influences chosen came from far and wide—eclectic, some of this and some of that. Specific definitions of styles based on eclecticism are somewhat hard to pin down. When interpreted by patrons and builders not formally trained, the results vary even more. In small towns like Madison there are usually few architect-designed buildings, and when they appear they are often commercial, governmental, or ecclesiastical but rarely residential. Absent budgets for architects, clients again turned to pattern books for styles and floor plans, just as they had for hundreds of years before and continue to do today.

After the catastrophic downtown fire of 1869, Madison's commercial area was rebuilt into a mostly red-brick, Italianate village of storefronts and warehouses. It was a building surge of necessity, and it ran counter to the na-

Top: The Hunter house on South Main Street (c. 1883) is picturesque in its asymmetry and mixture of Queen Anne and Second Empire styles. Below left: The Martin Richter house on South Main is an 1830 house renovated several times, beginning in 1885, with Italianate and Queen Anne details. Below right: The Peter Walton house (c. 1889) on South Main incorporates Queen Anne and Colonial Revival features in its design and decorative elements.

POSTBELLUM ARCHITECTURE

tionwide economic panic of 1873 and the prevailing economic malaise hovering over the Reconstruction South. Otherwise, construction of all types in Madison during the 1860s and '70s was limited mostly to additions and remodeling, adding a room here or a porch there, as finances permitted.

As the economy improved, bolstered again by cotton, buildings and homes began to go up in a variety of Victorian era styles. Certainly the most prevalent was the so-called Folk Victorian, where scroll-sawn trim and turned or chamfered porch posts were applied to middle-class vernacular houses. Fashionable trim was turned out by the local Madison Variety Works, where modern saws and jigs could accomplish in minutes (and at an affordable price) what would take days for an artisan with hand tools.

Initially there was a continuation of the Italianate and Gothic Revival used before the war. Then the Romanesque Revival found expression at the Madison Graded School, and Queen Anne and Eastlake combinations appeared around the town in brackets and arches, turrets and balconies, as new constructions began to break away from the symmetrical formality of the antebellum styles.

Before long, however, there was a gradual but predictable return to the classical and familiar designs of antiquity. In 1893 the World's Columbian Exposition at Chicago celebrated four hundred years since the discovery of America in 1492. It was from those exposition fairgrounds in Chicago, designed by New York Beaux Arts neoclassicists McKim, Mead, and White and others, that the nationwide classical revival came at the turn of the century. The 1905 domed and columned Morgan County Courthouse, designed by James W. Golucke, is an example of this trend. Then there are the outstanding houses around Madison built in the Neoclassical Revival style to compliment the town's many white-columned antebellum Greek Revival mansions from a half-century before.

The period from 1875 to 1917 has been called the American Renaissance, when the nation grew mightily between catastrophic wars and debilitating depressions. In the postwar, post-slavery South, economic recovery was also accompanied by a new social order, and adjustment for many was not easy. It is no wonder that those who needed some sort of comfort and reassurance found it in the designs of their ancestors.

Below left: The Foster house on Academy Street is a reworking of an early nineteenth-century house with substantial additions and a Colonial Revival makeover. Below right: The Joshua Hill house (c. 1842) on Old Post Road was renovated in 1917 with a massive Neoclassical Revival portico. Opposite: The Beaux Arts Morgan County Courthouse (1905–7) on East Jefferson Street.

POSTBELLUM ARCHITECTURE

Rose Cottage
179 East Jefferson Street
1891

Adeline Rose built this house as her home in 1891 in West Madison near the Georgia Railroad right of way. She was born in September 1864, in slavery, as the Civil War was ending, and died March 10, 1959, after living in the house for sixty-eight years. In 1996 the city of Madison moved her cottage to this location next to the Rogers House Museum and the county courthouse to restore it as a house museum. It honors Mrs. Rose, a widow with two children, who supported her family taking in laundry at fifty cents a load. For a time she did washing and ironing for the boarders at the Hardy House, which was owned and managed by the mother of famous comedic actor Oliver Hardy.

This Folk Victorian style one-story house, the Rose Cottage Mu-

Opposite: Victorian style was made affordable for even small cottages by the Madison Variety Works. Above: A photographic portrait of Adeline Rose hangs above the bedroom mantel. Right: In the kitchen are period appliances and utensils.

seum, represents changing aspects of the postbellum society of Madison, from slavery through Reconstruction, the late nineteenth-century New South era, and the twentieth century. Like the Rogers house next door, it is open daily and managed by the Morgan County Historical Society in cooperation with the Morgan County Board of Commissioners.

Atkinson-Rhodes House
408 South Main
1893; second story added in 1900

Millard Fillmore Atkinson, a partner in the Madison Variety Works, built a one-story house on this site for his marriage to Junie Baynes, December 14, 1893. In 1900 he enlarged it by adding the second story, a demonstration of the Queen Anne style that the saws of his family's Variety Works could accomplish. All over town there are many other examples of this work from the late Victorian era.

The Atkinsons' daughter Martha was born in this house in 1899 and lived here until she died in 1983. Martha was the music teacher in the grade school next door (now the cultural center) for thirty-seven years. In 1929, she married Paul Rhodes, who came to Madison in 1916 as a partner in Rhodes-Smith, a dry-goods store on Main Street at Jefferson. In 1956 Martha's sister, Mrs. Warren (Helen Atkinson)

Opposite: The lacy arches of the Atkinson-Rhodes porches add a romantic accent to the frame house. The scroll-sawn details were applied to the doll house (opposite, bottom) and to the interiors (above). The entrance/stair hall is a fantastical combination of shapes and colors.

The interior trim of the Atkinson-Rhodes house showed off the precision designs made possible by Atkinson's Madison Variety Works. Opposite: Dining room from hall. Above: Parlor.

DeBeaugrine, returned to live at home with her daughter, Martha. Helen also taught music in this house, and her other daughter, June DeBeaugrine Harrell, a landscape architect trained at the University of Georgia under Dr. Hubert Bond Owens, now resides in this Atkinson family home. June is active in Madison's civic affairs and has served on the Madison Historic Preservation Commission.

The elaborately decorated lacy-latticework front porch extends from the projecting front gable, around the north side of the house. At the second-story level, a one-bay arched porch is part of the exotic Queen Anne style decoration. A garden playhouse, built for the Atkinsons' daughters at the family's Variety Works and delivered by a mule-drawn wagon, reflects the Victorian period decorative motifs of the house. William Chapman's Madison *Manual* describes the entire ensemble as Folk Victorian.

Shaw-Erwin House
364 Porter Street
c. 1847; enlarged and remodeled in 1897

Alfred Shaw, a merchant who owned a furniture store in downtown Madison, built this house about 1847, and its history is tied to the history and character of Madison in several ways—the devastating fire of 1869, the prominent place of education in town, and the role of Methodism among the citizenry. The fire that consumed almost all the commercial district in 1869 is thought to have started in Shaw's store. Alfred's son, Horace Thompson Shaw II, a Confederate veteran who died in 1870 at age thirty, owned the house next door. Horace's widow, Sarah, operated a school for girls at this home for many years. The Shaws' oldest child, Simeon, was a Methodist minister and a missionary to Japan who later moved to Texas and, according to family tradition, was involved with the founding of Southern Methodist University. The second son, Horace III, once owned a majority interest in the Bank of Madison and was instrumental in building the First United Methodist Church in 1914.

During the Victorian era, the house was renovated into a fashionable Queen Anne style villa, which it remains today. After the Shaws and other owners, it was the home in the 1970s and '80s of Dr. Charles Lower and noted local artist Martha Lower.

Across the street is the Furlow house, c. 1850,

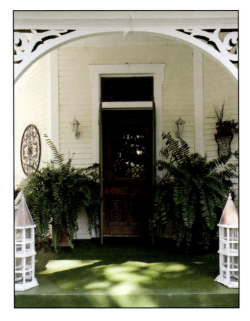

158 *Madison: A Classic Southern Town*

Opposite: The house was remodeled with fashionable Queen Anne details in 1897. Above and right: The Erwins enjoy their backyard landscape of gardens and outbuildings from their screened porch.

known as the Boat House. In the 1980s and early '90s, it became a bed-and-breakfast operated by Ron and Rhonda Erwin, who moved across Porter Street in the mid '90s to the Shaw-Lower house. The Erwins have made their spacious backyard into a veritable outdoor museum of gardens and outbuildings; a barn displays an antique pickup truck.

Ron Erwin says, "I wasn't born in Madison, but I got here as fast as I could!" He and Rhonda have operated a well-known shop, Laughing Moon, in the 1880s Belmont Hotel Building on South Main Street since 1984, when they moved to Madison. They also have a home furnishings shop in an old commercial building on West Jefferson Street—In High Cotton, an appropriate name indeed for Madison, because the fortunes of the town often rose and fell with the price of this commodity.

Magnolia House
356 South Main Street
c. 1839; remodeled in 1898

In the 1890s this house took on the form and style it has today. It started as a Greek Revival two-room cottage with a central chimney and a dependent kitchen; the downstairs bedroom, bath, and kitchen descend from that 1839 cottage. On December 17, 1853, Dr. William Burr, a dentist, purchased the property to serve as his residence and dental office.

The family to own the property longest after Dr. Burr was that of Colonel Edward W. Butler. The son of a Baptist minister, Butler was an attorney and teacher. He and then his widow owned the house from 1890 until 1937, when it was bought by the Butlers' daughter, Virginia Butler Nicholson, who bequeathed it to her husband, Dr. J. H. Nicholson. Nicholson left it to his second wife, Gladys Wallace, in whose family it remained until 1980.

George "Ed" and Louise Hannah purchased the house in 1993 and decided to renovate it as their home. They retained Atlanta preservation architect Lane Greene to help. Greene determined how the structure evolved from 1839, noting that it became two stories in the 1890s Queen Anne phase, which remains today. Research uncovered a quotation from the *Madisonian* dated July 22, 1898: "The home of E. W. Butler is nearing completion and will be one of the handsomest and most convenient in the city when it is finished." After World War II the house was divided into four apartments, becoming the Magnolias.

The Hannahs call it Magnolia House, and Louise Hannah keeps careful files on the history of their home, including a detailed printed leaflet describing the renovation/restoration, begun in 1994, that she and Ed accomplished with the help of Lane Greene. In her leaflet she writes, "Attention has been given to retaining the architectural integrity of the house while making it Twentieth Century comfortable." Louise Hannah serves on the Madison Historic Preservation Commission.

The Hannahs' Magnolia House stands between the Baptist and Presbyterian churches on South Main Street in a park-like setting.

The Hannahs have furnished their home to compliment the fashions of the 1898 remodeling. Above: View from hall into living room. Below: The master bedroom.

Left and above: The dining room is a riot of patterns, curves, and textures. Below: The detached kitchen is now a guest house.

MAGNOLIA HOUSE   163

La Flora
601 Old Post Road
1895

In 1895, when A. K. Bell built this house, Georgia Commissioner of Agriculture R. T. Nesbitt wrote a book in which he described Madison as "one of the most beautiful towns in Georgia. . . . noted for the taste of ladies in the cultivation of flowers in their beautiful flower yards." He might have been talking about La Flora, then and now.

Current owners Jack and Nancy Miles, who met in college in Indiana, discovered Madison in the spring of 1992 when Nancy came to tour the town because of its "beautiful flower yards." She was quoted in the *Madisonian*, April 16, 1998: "I fell in love with Madison. Jack and I have always collected antiques and admired historic homes."

Opposite: Their gardens confirm that Jack and Nancy Miles have taken to the implied responsibility of owning a place called La Flora, the namesake of a prominent Madison garden club. Above: Living room with views into hall and dining room.

The Mileses are La Flora's fifth owners since it was built in a combination of the Queen Anne and Colonial Revival styles. They bought it from Mr. and Mrs. Charles Cartwright Jr. in 1992 and began the renovation. They especially admired the original Queen Anne style parquet ceilings that were milled at the Madison Variety Works. They have carefully restored and furnished the place with period pieces so that it is a step back in time, which pleases them no end. Jack has served on the Historic Preservation Commission, and Nancy has been the president of the Magnolia Garden Club.

The house is named La Flora because it was associated with the

The parquet Queen Anne ceilings were milled at the Madison Variety Works. Above: Dining room. Below: Entrance/stair hall. Opposite, bottom left: Family room. Opposite, top and bottom right: The c. 1890 guest house, a Madison tradition.

rejuvenation of the Madison Garden Club, founded in 1893, the second garden club in Georgia and the first in Madison. The club stopped activities in 1920, but Mrs. Harris Richard initiated its reorganization in this house and renamed the club La Flora. Her daughter was Dorothy Richard Baldwin, whose husband's family owned the house for many years.

On the rear of the property facing West Walton Street is a quaint two-room cottage where Jack Miles's mother lived for years. A tradition in Madison is the rescue, relocation, and restoration of one- and two-room houses as guest cottages, a practice made possible by the unusually spacious lot sizes in the town's residential neighborhoods.

LA FLORA

White-Lyle Cottage
254 Pine Street
c. 1900

This pyramidal-roof Queen Anne cottage was originally part of a larger property on East Washington Street owned by the Speed and Bearden families. Built around the turn of the twentieth century, the house represents a popular middle-class form used throughout the region for decades after the Civil War. Basically a four-square, central hall plan with a steeply pitched roof, its variations were usually distinguished by the type of trim on the eaves and porches.

Miss Nellie Booth acquired this house in 1908 and transferred it in 1933 to W. E. White, whose family sold it to the current owners, Bob and Penny Lyle, in 1991. The Lyles were residing in Naples, Florida, but Penny had lived in Madison when her children were young and held a lingering fondness for the town and its piedmont environment.

The Lyles engaged several local talents to transform the place into a tranquil retreat that became their permanent home. Leonard Wallace, an interior designer with deep Morgan County roots, helped the Lyles understand how the earliest residents might have furnished this house. Landscape architects

Opposite, top: The Lyles' cottage is distinguished by Victorian trim and a pyramidal roof. Opposite, bottom: The sun and screened porches are part of a 2006 addition designed by Joe Smith. Above: Living room. Below: Smith designed the guest cottage modeled on Madison precedents.

Richard Simpson and Rick Crown, whose art and skills are evident throughout the county, established an inviting woodland back garden.

Award-winning improvements were designed by Joseph Smith, a local historic preservation–minded architect with Hall Smith Office: in 2006 a rear addition was honored by the Madison Historic Preservation Commission for appropriateness, and in 2009 a small guest cottage based on an 1880s tenant house on the Barnett-Stokes property was recognized for being an especially "compatible new outbuilding." Madison is fortunate the Lyles decided to come home from Florida to live and contribute ably to the town's preservation movement.

Morgan County African-American Museum
156 Academy Street
c. 1904

John Wesley Moore's house was moved to Academy Street to be restored and furnished with displays and collections relating to the life of African Americans in Morgan County and the South.

John Wesley Moore was born in the last years of slavery, in January 1862, and lived to be forty-six years old, dying in 1908. He married Dora Gordon on November 21, 1881, and they lived in a tenant house on land owned by a white farmer, James A. Fannin. The African American couple's first child was born in October 1883. On April 10, 1890, Wesley Moore bought five acres of land; by then he and his wife had four children. On October 31, 1899, Fannin deeded Moore forty-one acres of land, "for five dollars in consideration of the service he has given me."

After Moore died in 1908, his widow inherited his land and other property. She lived until 1932 in this house, which has since been moved from Bethany Road in the county to a vacant lot at 156 Academy Street to be the Morgan County African-American Museum. According to Marshall "Woody" Williams, the longtime county archivist, "The original structure could not have been earlier than 1890, when Moore bought the property, and could have been as late as 1910, when Dora Moore owned the property free and clear." Williams speculates that "a construction date of about 1904 would be indicated," but it could be as early as 1895.

The simple, Victorian period, one-story frame house, donated by Reverend Alfred Murray and now furnished and filled with exhibits, was dedicated as a museum July 11, 1993, on the property between the Calvary Baptist Church and the Round Bowl Spring Park. Modern improvements were made, with the renovation work led by Robert Boswell and Robert Lewis, who donated their services to the cause.

Godfrey-Hunt House
586 Academy Street
1875; expanded and renovated in 1895; renovated in 1922

This house, built in 1875 on the site of a much earlier structure, is currently the home of Mr. and Mrs. Lowry Weyman Hunt Jr., Whitey and Lyn. They moved here, Whitey's family home, from Elliot Acres (the Rogers-Shields house) on North Main, Lyn's family home in Madison. Built by Whitey's great-great-grandparents with a two-over-two-room, side-hall plan, this house is one of only a few in Madison owned by the same family since its construction.

The house was built by Dr. James Ervin Godfrey (1834–87), a surgeon in the Confederate Army, a state senator, a plantation owner, and, in 1879, founder of Godfrey's Warehouse. Now operated by his

Architect Neel Reid in 1922 renovated the house inside and out in the Colonial Revival style for the Godfreys. Above: The front porch is a pleasant outdoor living room for entertaining and relaxing.

Hunt descendants in its original location, Godfrey's is the oldest business in Madison continuously owned by the same family. About 1895 Mrs. Godfrey, the former Mary Perkins Walton (1839–1921), expanded and renovated the house, adding two rooms, one upstairs and one down, and a new, one-story, wraparound front porch. Her family, the Peter Wyche Waltons, had lived next door, where she and the doctor were married in 1859 and where Oak House now stands.

A few years after his mother's death in 1920, Captain J. E. Godfrey and his wife hired Atlanta architect Neel Reid (1885–1926), a leading early-twentieth-century Georgia classicist, who remodeled the house in his characteristic Colonial Revival style. Reid's only Madison project (Hentz, Reid & Adler job number 465), it demonstrates the archi-

The Hunts have appointed their home with items they treasure from their families. Opposite: A screen of Ionic columns on pedestals separates the living room from the entrance hall. Above: Dining room. Right: Carved mantel in the study.

tect's subtle touches of Ionic classicism, inside and out.

Whitey's parents, Caroline Candler and Lowry Weyman Hunt Sr., bought the house from her great-uncle's wife's estate in 1952. Caroline Hunt was a well-respected local historian who also connects to Honeymoon, on Eatonton Road. The Hunts remodeled the house in 1969, and in 2003, Lyn and Whitey moved in and began their renovations, hiring preservation architect W. Lane Greene, of Atlanta. They have furnished the home with pieces from both their families, and in these elegant surroundings, a new generation of Godfrey descendants continues to treasure family traditions.

Oak House
617 Dixie Avenue
1897–98; additions and renovations 1992–94

Lee Trammell built this New South Neoclassical Revival house in 1897–98 on the site of a Walton family antebellum home that apparently burned not long before. Trammell was the grandfather of Floyd Newton Jr., of Boxwood, who remembers his mother, Mary Trammell Newton, stating that she had established the boxwood gardens here. Mrs. Newton died in 1984, and in 1986 the house was sold to Mr. and Mrs. Thomas E. DuPree Jr.

Tom DuPree, a native of Macon, chose Madison as his new hometown because of its historic character, physical beauty, and community spirit. In fact, DuPree is mayor of Madison in its bicentennial year of 2009. In 1992 DuPree commissioned Hansen Architects to begin a major renovation with new additions in keeping with the house's original neoclassical architecture, exemplified by the four-columned portico in the rare Roman Composite order. The grand staircase in the entrance/stair hall has intricate oak-leaf carvings on the newel post, thus the name Oak House.

A Georgia native, interior designer Charles R. Davis Jr. of New York became the decorative arts consultant for the project. lacy-Champion, a renowned North Georgia textile firm that provided the carpet for President Jimmy Carter's Oval Office, made most of the

Oak House sits back from Dixie Avenue in a park of trees and hedges. Below: The skylight above bridges connecting the original house to the addition. Right: The entrance stair hall and carved staircase.

Oak House rugs especially designed for the project. The stained-glass skylights were fabricated by Rambusch, a family-owned New Jersey company providing architectural designs for lighting, art metal, and skylights for installations as diverse as the New York Stock Exchange, the Statue of Liberty torch, and the Ronald McDonald House on Wall Street.

Upon the completion of the Oak House project in 1994, an award from the Georgia Trust for Historic Preservation recognized the excellence of the restoration and renovation of this estate at Dixie Avenue and Walton Street in the Madison Historic District.

Opposite: The transverse hall opens onto a large boxwood garden framed by attractive fencing and hedges. Above and left: The garden side of the house clearly shows how the addition was carefully designed to the scale and style of the neoclassical original.

180   *Madison: A Classic Southern Town*

Opposite, top and far left: The double parlors are separated by a columned arch. Opposite center: A curved wall and decorative niche at the entrance to second-floor bedrooms. Opposite, near and above: The formal dining room is in the 1992 addition.

Porter-Fitzpatrick House
507 South Main Street
c. 1850; renovations in 1901

William Chapman includes this house in his *Madison Historic Preservation Manual* in a discussion of the Neoclassical Revival: "Many seemingly antebellum houses in Madison are classical revival reworkings of earlier buildings. The Porter-Fitzpatrick-Kelley House was built in the 1850s and remodeled in 1901" (page 71). The builders were John W. Porter and his wife, Mary Wade. In 1901 Henry Harris Fitzpatrick made extensive renovations. He made a new entrance façade on South Main Street, bringing it up to date with turn-of-the-century neoclassicism that had become fashionable because of the World's Columbian Exposition (or Fair) of 1893, the so-called White City on the Lake Michigan shore. The six giant Roman composite columns set on high pedestals are a monumental feature.

Opposite: The Main Street side of the Beckers' house was renovated with a Neoclassical Revival portico in 1901.
Above: The entrance/stair hall.

The Fitzpatricks owned the house until the early 1970s, when it began to pass through several owners. They included the Roleys, who made significant renovations, and the Frank Kelleys, who redecorated the interior and built an addition on the Old Post Road side, which the house had originally faced prior to the Fitzpatrick renovation in 1901.

The current owners, Don and Rosie Becker, purchased the house in 1999 and began some major changes the next year, principally adding a garage and expanding the summer kitchen into a guest house. In 2008, they continued the tradition of renovations to this house, with interior remodeling and structural stabilization, completing the arduous undertaking during the bicentennial year.

The house underwent extensive Neoclassical Revival stylistic changes throughout in 1901. The Beckers redecorated the interiors in 2009 with a fresh palette of colors and fabrics. Opposite, top: The living room. Opposite, far left: The original front of the house faced Old Post Road. Opposite, near: The front portico rises above the deep evergreen garden. Above: Dining room. Left: Kitchen.

Poullain Heights
766 East Avenue
1905

This property, with its expansive grounds, canopy of large trees, and allée of oaks, is located only about a half-mile from the Madison town square, but it evokes the notion of a country estate. The two-story Neoclassical Revival house was built in 1905 by two sisters from the Poullain family—Sara (Mrs. Robert Harris Campbell) and Florida (Mrs. Charles L. C. Thomas)—and the property became known as Poullain Heights.

Today it is surrounded by a small-town suburban landscape, but it is part of the unlikely intersection of two remarkable families, each descended from prominent surgeons in the Revolutionary War. Dr. Lancelot Johnston was commissioned as a surgeon in the Ninth North Carolina Continental Regiment and is reputed to have been a physician to General George Washington. Antoine Poullain came to America from France as the surgeon to the Marquis de Lafayette. Their sons became famously wealthy plantation owners—Lancelot Johnston, who also prospered as an inventor in Madison, and Dr. Thomas Poul-

Opposite: The neoclassical house at the end of an allée of oaks conjures images of bygone times. Above left: Parlor and dining room. The inn is decorated to encourage guests to feel at home and comfortable. Above right: Dining room. Below: Stair hall.

lain, who owned a successful cotton mill in Greene County. Their children, Antoine Poullain and Elizabeth Jones Johnston, were married in 1842 at Lancelot Johnston's home, Snow Hill, which was a much-admired showplace in antebellum Madison.

Snow Hill (which burned in 1962) eventually passed to Elizabeth and Antoine Poullain, and two of their daughters, Sara and Florida, built Poullain Heights on family property across the road. For many years it was the home of Miss Florida Campbell Prior, a Poullain descendant who lived here with her mother and inherited the place in 1958. In the 1980s it was owned by the Baron and Baroness Werner von Hanstein and then, in the early 1990s, by Ed and Louise Hannah, now of Magnolia House.

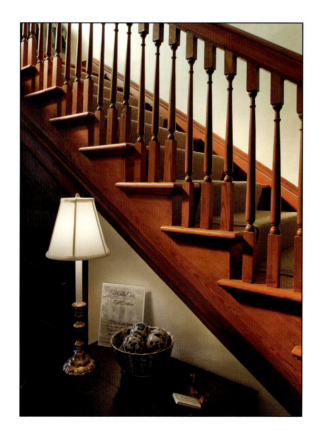

In more recent years, this Neoclassical Revival in-town villa has become the Madison Oaks Inn, a comfortable bed-and-breakfast owned and managed by Roger and Dianne Simmons. Four well-appointed bedrooms, each with a private bath, are named for former resident-owners of Poullain Heights: the Sisters' Room, the Thomas Room, the Baron's Room, and the Hannah Room. The handsome portico with four white Ionic columns, framed by leafy arches at the end of the oak allée, brings to mind iconic visions of the antebellum South. Though the house was built forty years after the end of the Civil War, its image is intended to reflect a connection to the lives and lifestyles of times past but not forgotten.

The Modern Era
Styles and Trends, Historic Preservation, and Revitalization

To study properly the architectural history of Madison's modern era, we must examine not only the styles that have left their imprints on the town but also the strong local commitment to historic preservation and the disparate societal trends toward urban revitalization and country retreats.

The modern era in architecture in the United States originated in Chicago at the same time as the neoclassical World's Columbian Exposition held there in 1893. Two schools of design developed, one precedent- and history-oriented, led by the New York firm of McKim, Mead, and White; and the other, "form follows function," growing out of the Arts and Crafts movement, led by Chicago architects Louis Henri Sullivan and Frank Lloyd Wright. The influence of both schools spread throughout the nation. Madison experienced both neoclassicism and Arts and Crafts experimentation during this era and continues to do so, especially in the historic preservation aspect of historical precedent architecture in the modern era.

The historical precedent orientation of neoclassicism eventually led to the restoration and historic preservation movement. The Colo-

The notion of attractive affordable housing evolved from late-eighteenth-century Folk Victorian details on cottages to the twentieth-century Craftsman and Bungalow styles found throughout Madison. Illustrated below on South Main Street are the Pool (c. 1906) and Vason (c. 1910), houses Opposite bottom: Saye house (c. 1915).

188 *Madison: A Classic Southern Town*

Rebuilding downtown Madison in the 1860s and '70s after the terrible fire was perhaps more dramatic, but no more important than the revitalization begun there a century later. The result has been a village center of shops, offices, restaurants, galleries, public buildings, and residences, and the charm and vitality is genuine. Opposite: Main Street at Washington. Above, from left: The 1887 city hall (now the Chamber of Commerce), the 1905–7 courthouse, and the 1931–32 post office illustrate styles from Italianate through Beaux Arts to Colonial Revival.

nial Williamsburg restoration in Virginia during the 1920s and '30s was a leading popular influence in the United States, resulting in such things as the incorporation of the National Trust for Historic Preservation in 1949. The restoration of antique buildings, of whatever style and period, has eventually been one characteristic of the modern era. The passage of the Historic Preservation Act of 1966 led to the National Register of Historic Places. Madison/Morgan County's Cedar Lane Farm was registered in 1971 and Bonar Hall in 1972. The first version of the Madison Historic District was included in the register in 1974 and expanded in 1989.

Madison played a prominent role in a regional spring historic preservation conference held by the Georgia Trust for Historic Preservation, a statewide organization founded in 1973. The conference of May 1974 in Madison, Eatonton, and Greensboro was the first annual meeting of the Georgia Trust after its incorporation. The trust chose Madison to be involved because it was already a state leader in historic preservation, and Madison's large treasury of antebellum domestic architecture, well-preserved but living landmark homes such as Boxwood and Hilltop, was well known.

When we discuss Madison's architecture in the modern era, therefore, we are concerned with more than the styles made popular in the last one hundred years. There are Madison examples of practically every twentieth-century style, from the Colonial and classical revivals and Beaux Arts classicism to Craftsman and Tudor cottages and Prairie interpretations. There are grand mansions which look like their antebellum antecedents and ver-

THE MODERN ERA

The twenty-first-century subdivisions at Valley Farm (above) and Candler Lane (below) serve a clientele seeking traditional home designs with modern amenities. Each was conceived to fit within the context of Madison architecture and lifestyles.

nacular cottages which appear timeless. Each is valued in Madison for its own character, and contemporary subdivisions such as Valley Farm and and Candler Lane feature designs that encourage memories of those foundation designs from the past.

Madison's modern era has also seen the reemergence of an ancient cultural form—the country villa. Made popular in Italy during the Renaissance, the villa is more about concept than architecture and was an evolution of an even earlier theme, the country house as a retreat or refuge from hectic and crowded urban life.

In the antebellum piedmont the opposite was true, however, as plantation owners preferred the comforts and amenities of town or city life to the relative social, intellectual, and educational isolation of their large farms. Now the trend has been reversed again, as urbanites from Atlanta and other cities seek not only the refuge of the country, but the stimulation as well, in the form of farming and gardening, recreation, and wildlife.

Above: The Town Park pavilion and the James Madison Inn are part of an ongoing effort to attract activity and commerce to downtown Madison. Below: Country villas like Camp Boxwoods offer refuge and recreation for those seeking relief from city life.

At the same time, other contemporary Madisonians are taking a different tack, adapting commercial buildings downtown for residential living space. Following the lead of the national Main Street program, Madison in the early 1980s became one of the first communities in the Georgia Main Street movement, despite having fewer than four thousand residents. The appointment of a downtown development authority served to encourage the stabilization and then revitalization of the city core. The traditional shopkeeper approach to downtown living, with owners living above their stores, took hold.

Madison at its two-century mark is a model for others to follow: its center holds a lively variety of shops, offices, restaurants, and galleries. There is a new downtown hotel and conference center, and most recently, a new central event facility and green space—Town Park.

Madison was honored in 2004 as one of Georgia's first Preserve America communities for its preservation ethic and for building its future on the best of its past. Madison is now as much modern era as it once was antebellum, Victorian, and neoclassical.

The Modern Era

BOOKHAVEN
606 NORTH MAIN STREET
c. 1909

Bookhaven sits embowered above North Main in a street-lined forest. Below: The parlor houses books and items of all descriptions relating to Charles Dickens and his works, especially A Christmas Carol. *Opposite: The hall has a display of photographs and autographs from members of the Algonquin Round Table. The dining room features a collection of volumes and images about Abraham Lincoln.*

Louis Carbine, one of the founders of the Farmers Hardware Store, built this house in 1909. After there was a fire in 1938 it was turned into a duplex, where many young couples started housekeeping, among them Adelaide and Graham Ponder, who went on to become the editor and publisher of the *Madisonian*. In 1975 the house became a single-family dwelling again.

Steve and Rita Schaefer bought the house in 1980 and began the renovation to make it into their home. They named it Bookhaven, because it had been the home of three librarians. For many years Steve was the director of Madison's Uncle Remus Regional Library, which serves four counties. He and Rita met at Florida State University when they were studying library science. They married and moved to Madison from Vidalia, Georgia, where Steve had been involved in setting up the well-known Jack Ladson Genealogical and Historical Collection in the Vidalia Library. Rita was Madison's elementary school librarian for many years.

Bookhaven is the Schaefers' residence, but it also houses their own book and memorabilia collection. Along with an estimated six thousand volumes are a room devoted to Abraham Lincoln, another to Sherlock Holmes, and another to Charles Dickens, especially his classic, *A Christmas Carol*. The Schaefers also collect Algonquin Round Table memorabilia from the 1920s when the midtown Manhattan Algonquin Hotel was the venue for famous gatherings of witty literary and theatrical notables such as Alexander Woollcott, Dorothy Parker, and the *New Yorker* magazine founding editor, Harold Ross. Manhattan may be a long way from Madison, but not so far at Bookhaven.

Carter-McManus House
667 Billups Avenue
1915

This 1915 Craftsman bungalow, called Echo House by its current owner, was built for the late Dr. Dan Carter from plans by Leila Ross Wilburn of Decatur, a pioneer Georgia woman architect, who spread her lasting influence throughout the Southeast. Today her work can be seen in many enduring residential landmarks in Atlanta's Druid Hills and Candler Park.

The very existence of a female architect signified a new era in American design, of which the Craftsman/Arts and Crafts movement was a significant part. The beauty of the Craftsman ethic is best expressed in Echo House by the sweeping expanse of the living room, with its twin fireplaces, built-in bookcases, and ten-foot box-beam ceilings—all of it flooded with light and open through matching French doors to a fifty-foot brick and tile terrace.

Although it was built from Wilburn's plan book, there is evidence that she consulted with Dr. Carter and made some changes to the design. In a copy of the plan book there are notations to that effect, including one apparently in Ms. Wilburn's hand: "'Rather odd, but somehow I like it.' That is what the owner's friends say about it."

This architecturally significant Madison house, which sits on a one-acre lot, was purchased in October 1999 by James J. McManus, a retired national news correspondent, and also is home to Ms. Leticia Simbach, of Honolulu, Hawaii. McManus has renovated this 3,200-square-foot home during his almost ten-year ownership and relates that "Visitors recall this spacious home as the scene of weddings and at least one festive Morgan County High School senior prom."

From a lifetime of travel in his occupation, Jim has visited cities and towns all over the world. When he decided to settle into retirement, he took out a map of Georgia and a attached a length of string representing a sixty-mile radius to a pin stuck "in the center of [Atlanta's] Fulton County Stadium." He decided to visit every town of a certain minimum size within that scribed circle, but when he got to Madison, he stopped searching. A decade later he continues to appreciate the proximity of Atlanta's amenities while immersing himself in the pace and pleasures of small-town living.

Jim McManus has shown obvious respect for the Craftsman style in his choices of colors and patterns. Above and below left: The long living room has a facing pair of fireplaces and French doors. Below right: The entrance porch is an outdoor extension of the living room.

Douglas-McDowell House
826 South Main Street
1917

Larry and Dell Morgan of Madison and Douglasville, Georgia, have been renovating this landmark of the modern era since June 2004 when they bought it with the idea of eventually making it their primary residence. They said they had "admired the old homes of Madison" for many years. They bought the house from Paulette and Lloyd Long, who now live in the Ice House lofts in downtown Madison.

Local historians have described this house as a "vernacular example of the

Prairie style house made popular in the early twentieth century by architect Frank Lloyd Wright" (*Madison, Georgia, An Architectural Guide*, page 80). It was built in 1917 by Mr. and Mrs. Tillman Douglas on the site of an earlier house that had burned. Later, when R. Frank McDowell, an owner of the McDowell Grocery Company, had the house renovated, he found signed inscriptions, one behind an upstairs mantelpiece, "This house was built in 1917," and another in a closet, "Rufus

Madison: A Classic Southern Town

This early twentieth-century southern suburban house had contemporary influences, but it was built around a classic form and plan and retained decorative elements from the Neoclassical Revival. Opposite, bottom, and above: The living room, entrance/stair hall, and dining room.

Williams painted this house inside and out, August 1917."

The Prairie style came at the same time as the Craftsman and Colonial Revival styles, and this house is essentially a synthesis of a Frank Lloyd Wright horizontal Midwest approach with the more conventional Colonial Revival classicism of the sort J. Neel Reid of Atlanta designed for many Georgia neighborhoods around 1920. The interior is especially reminiscent of Reid's 1920s classicism. Both Wright and Reid often used open front porch terraces with pier pedestals for displaying plant boxes, helping to relate the house to the landscape.

Burney-Ponder-Rushing House
912 Dixie Avenue
c. 1875; renovated in 1948, 1999, and 2003

Above: The façade was renovated in 1948. Below: A view into the rear garden. Opposite: From the entrance hall into the living room.

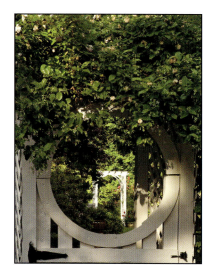

Bill and Bobbie Rushing moved to Madison from Atlanta in 1999 after seeing this triple-gabled cottage the previous year and deciding it would be perfect for the next phase of their lives. The 1875 house originally had only one front gable, but it was renovated in 1948 by Paul and Mary Ponder, the owners prior to the Rushings. They have renovated and restored it twice themselves, the second time after a fire partially gutted the house in 2002. At that point Bill and Bobbie had to face the difficult decision of whether to raze and start over or to begin again an expensive restoration. Fortunately for Madison they chose the latter, and they have continued the process of subtle renovations without taking away the historic character and charm of this 1870s Victorian era cottage.

Sincere preservationists who appreciate the context of history and tradition, the Rushings had once lived in an Atlanta house that *Gone with the Wind* novelist Margaret Mitchell and her husband John Marsh owned when Mitchell died in 1949. Bobbie, an interior designer, was once associated with Atlanta's well-known design firm led by Stan Topol, and her work on their own

Bobbie Rushing's interior design mixes historical themes and artifacts with contemporary colors, images, and schemes. Opposite: Library looking into dining room and kitchen. Above: Dining room. Below: Two views of the living room.

homes demonstrates her sensitivity to bringing old houses appropriately into the modern era. The dining-room mural of scenes from Irish history (which remind the Rushings of their own cottage in Ireland) and many cosmopolitan artistic aspects are distinctive design features of this personal reinterpretation, which has become a popular stop on Madison's tours of homes. The beautiful gardens, much loved by Mary Ponder, are now enhanced and carefully tended by Bill, presently chairman of the Madison Historic Preservation Commission.

Gilbert House
848 Dixie Avenue
2000–1

Dixie Avenue is a notable venue for antebellum domestic architecture in Madison. One of the great houses, Thurleston, is located directly across the street from the Gilbert house, which a civic-minded couple built in 2000–1 to blend with the historic setting. Indeed, the boxwood-bordered lawns of this house and Thurleston seem to merge into one landscape.

Judy and Bruce Gilbert are leading citizens, supporters of all things positive in Madison. Bruce has been mayor of Madison, and Judy, a native of nearby Eatonton, owns a real estate company. Together they encourage prosperity and cultural enrichment in this piedmont Georgia community, long known for the good life.

The architectural beauty of their home is

Opposite, top: The Gilberts' boxwood allée lines up with that of Thurleston across Dixie Avenue. Opposite, bottom: Displayed on the living room secretary is a photograph of the Ackerman-Little house, which burned. Above: The sitting room looks out onto a screened porch.

Above: Dining room and hall. Right: The entrance/stair hall opens onto a view over Dixie to Thurleston. Far right: The family room and kitchen combination are a modern notion in this traditional design.

Above: The spacious screened porch and shaded grounds give privacy and quiet.

largely due to Judy, who searched through her favorite books about historic Georgia architecture and architects to find exemplary models. Among books she specifically mentioned were *Edward Vason Jones, Architect, Connoisseur and Collector*, *J. Neel Reid, Architect*, and *Landmark Homes of Georgia, 1733–1983*. Of course she found classic models all around in Madison itself, among them the old Ackerman-Little house, a brick house constructed in the 1850s that stood on this very site until it burned in the 1960s. Judy displays a framed photograph of the old house in a place of honor in her well-appointed sitting room.

With research in hand and definite ideas in mind, Judy contacted architect Julie McClellan of nearby Watkinsville, and together they created a modern family home which fits comfortably within the architectural fabric of its traditional surroundings.

Samuel Hanson House
Davis Academy Road
c. 1816; renovated c. 1835; restored and renovated in 1985

For more than half a century, Morgan County has been a convenient and stylish countryside escape for affluent Atlantans seeking relief from city life. The pattern for this approach in the county was begun in the 1950s by the late decorative-arts aficionado Henry D. Green at Greenoaks Plantation, a Plain style farmhouse on Monticello Road built in 1815 that Green restored and saved from dereliction. About fifty years later, in 2005, this classic Plain style house from the same era became the personal retreat for one of Atlanta's leading interior designers, Toby West, and his partner Tom Hayes.

Located on a bend in Davis Academy Road, the old two-story frame house sits on a level building site with four front-yard oaks evidently planted about a century ago for shade. Typical of the early Plain style plans from about 1810–20, the house, built by Samuel Hanson, shows only a casual stab at symmetry, with a front door almost, but not quite,

Opposite: The asymmetry of the vernacular farmhouse has a charming lack of self-consciousness. Above: Dining room. Below: Breakfast room.

centered on the five-bay façade. (During the more self-conscious Greek Revival period, symmetry was almost always perfect.) The vertical Plain style profile, with tall outside brick chimneys and shed additions front and rear, reinforce the impression of authenticity.

Inside, the floor plan is an early hall-and-parlor type, with stairs rising from the original shed room to the rear, but Greek Revival details

imply an updated remodeling after the house was built, possibly in the late 1830s or '40s.

Once in an advanced state of dilapidation and being used for hay storage, the house was rescued in 1985 by Angela and Ken Lewis, who engaged Atlanta architect Norman D. Askins as their restoration consultant. The Lewises sold the place in 2001 to Jim and Cathy Fletcher, who made several changes, including adding the full-

SAMUEL HANSON HOUSE

Toby West's skills are evident in the easy way high-style art and accoutrements fit within the plank walls of this simple dwelling. Preceding pages: Living room. Above, top left: The den is an extended space off the stair hall. Above, left: Stair hall and den. Above, right: Upstairs hall. Opposite: Guest bedrooms.

length front porch, remodeling the kitchen, and relocating an early mantel salvaged from an old house in Madison. Since Toby and Tom have owned the property, they have enlarged a rear sunroom, built a new barn for their horses, and done extensive landscaping. All along, West has made this former farmhouse a design laboratory for his interior decorating skills and love of decorative arts, a well as his taste in architectural restoration.

Madison/Morgan has benefited greatly from its proximity to Atlanta and I-20 and the modern-era devotion to historic preservation. The town-and-country aspect continues to play an important supporting role in Madison's bicentennial story.

Willow Oak Farm
Bethany Road
c. 1867; enlarged and remodeled c. 1880s

In the 1840s this was a large cotton plantation—several thousand acres with many farm buildings and quarters for the slaves. Today it is a carefully maintained family retreat for recreation, sport, and wildlife management. It is still a significant size by most measures, but not as large as its expansive antebellum acreage. On the grounds are two primary dwellings, a house built about 1867 to replace an earlier dwelling and a modern but compatibly designed "gathering house" to accommodate family and guests.

A detached kitchen, perhaps dating from as early as 1845, stands on the east side of the postbellum farmhouse. Modifications and additions made by the Harriss family during the Victorian era gave the house its present configuration, with a gabled ell extended on the front and Italianate trim added. The first-floor window on the ell is decorated with a bracketed hood. Italianate features similar to these were added to houses throughout Madison and Morgan County in the 1870s and '80s, provided by the Madison Variety Works.

Mr. and Mrs. Frank Carter of Atlanta in 1968 purchased this property, the house and one hundred acres, to serve as a country home convenient to Atlanta for their growing family. Over time they have purchased additional acreage to provide room for hunting, fishing,

Willow Oak Farm provides many opportunities for outdoor recreation with golf greens, sporting clay stations, fishing ponds, and riding trails, but it also has vegetable gardens, timber tracts, and forage pastures for the pursuit of more traditional aspects of country living. Opposite: The c. 1867 farmhouse with detached kitchen. Above: The "gathering house" and vegetable garden. Below: Rear of original house.

sporting clays, horseback riding, and quiet relaxation. The Carters are a real-estate clan in Atlanta, but Wilson Carter, a Madison resident, says this is a "legacy property" that will remain in their family and not be traded as if it were part of their commercial operations. Simply put, this is now the Carter family's recreational homeplace. They named it Willow Oak Farm.

In the pattern of other Atlantans in the modern era, the Carters are investors in the area's past as well as its future. In May 2009, the Carter family shared Willow Oak Farm as part of the Madison Bicentennial Spring Tour of Homes.

STOKE FARM
SOWHATCHET ROAD
1994

Stoke Farm is located near Bostwick, north of Madison in rich farming country, and this is a 130-acre working farm, complete with cattle, wheat, orchards, and vegetable gardens. Historically, cotton was the cash crop in the area, of course, and there is still a Bostwick Cotton Gin, which is the only one still operating in Morgan County in 2009.

The house on the Stoke Farm property was built in 1994, designed for Mr. and Mrs. William A. Lobb of Atlanta by architect Steven Elmets, who was associated for a dozen years with one of the best-known American architects, I. M. Pei. It was Elmets's first commission on his own, and the Lobbs encouraged him to design a modern version of antebellum classicism appropriate for the rolling Georgia piedmont landscape. Elmets's façade design implies a version of Italianate neoclassicism (such as Boxwood in Madison), while the modern interior centers on a galleried atrium open to the belvedere and framed with four square columns. The white walls bounce the soft natural light from about and above and help

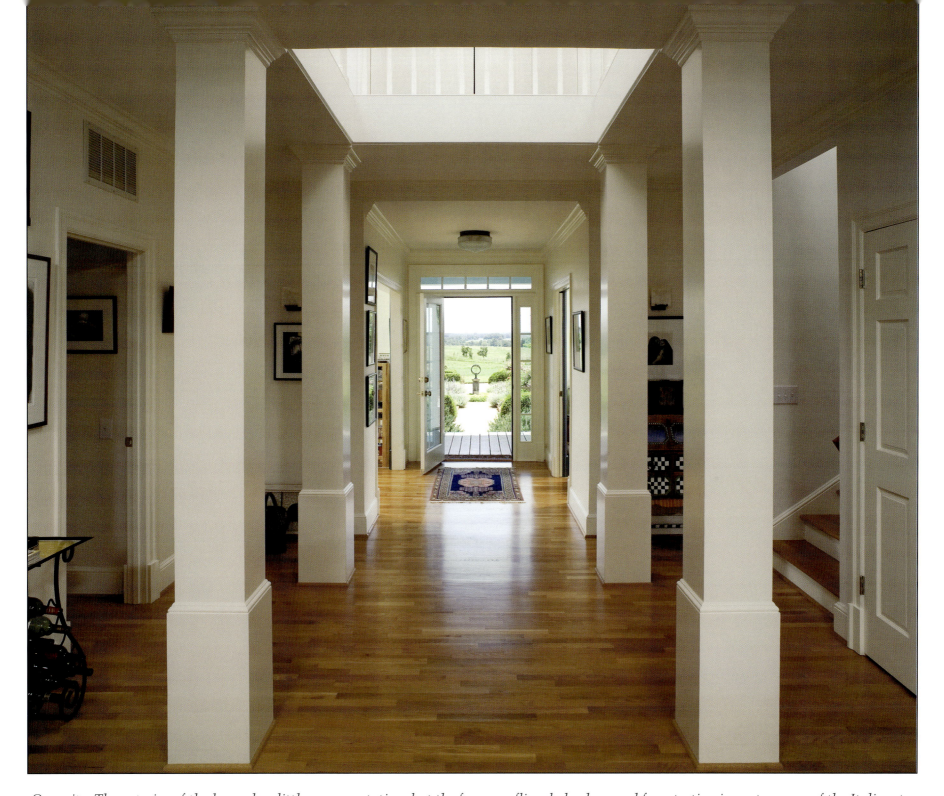

Opposite: The exterior of the house has little ornamentation, but the form, roofline, belvedere, and fenestration impart a sense of the Italianate. The parterre garden brings a formality to the open fields and vistas. Above: The central hall leads to a columned atrium.

illuminate the Lobbs' collection of fine photographs hanging throughout the house.

The formal boxwood parterre garden in the front is reminiscent of other Madison area gardens, some of which date back to the earliest days of the town and county. The entire composition satisfies the challenge presented by the Lobbs; it is a pleasing and intended combination of built and natural forms, of culture and agriculture, and it meets the classical standards of the farmhouse villa in a most appealing way. The name Stoke Farm derives from William Lobb's ancestral English roots at Stoke Point in Cornwall, England.

Camp Boxwoods
Athens Highway
2004

This is the country estate of Dan Belman and Randy Korando, successful interior designers and owners of Boxwoods Gardens & Gifts, a northwest Atlanta fixture since 1995. In 2002 they decided they needed an escape and a new project. The partners looked all over the state, "from the mountains to the sea," as we say in Georgia, and they settled on a one-story 1978 house overlooking a small lake on sixty acres north of Madison. Their property, called Camp Boxwoods for obvious reasons, is about sixty miles east of their Atlanta Buckhead shop in a county that also has a Buckhead, one which calls itself "the real Buckhead" because it was founded about forty years before the northwest Atlanta neighborhood by that name.

Thirteen months later, in 2004, they had transformed the simple lake house and property into a three-story Victorian-style garden spot. They considered it appropriate for an area with many fine old Victorian houses and gardens, which Madison/Morgan County assuredly has. (Madison already had a famous Italianate house, c. 1851, called Boxwood, so Camp Boxwoods seemed doubly appropriate.) The house

Left: The house is approached on a road winding through a forest and over a lake. Above: The rear herb and vegetable garden.

is a marvelous assemblage of salvaged architectural details, mostly from Victorian and Neoclassical Revival sources, pieced together in a wholly believable form.

The partners are both originally midwesterners, but they have felt at home in Morgan County, partly because there are many other part-time transplants here, including a group of friends calling themselves the East Atlanta Farmers, who maintain residences in both Atlanta and Morgan County. Randy and Dan have become interested in rescuing buildings and animals. The stables, created from an abandoned African American church on the site, is now home to four res-

Opposite and above: The living room and entrance hall were created with an assortment of architectural details procured around the world.

CAMP BOXWOODS

cued horses. They have adopted numerous sheep and some Scottish Highland cattle, a donkey, and llamas, not to mention ducks and assorted yard fowl.

The rambling, turreted house on a hill, the lush gardens, exotic animals, and the name, Camp Boxwoods, give a hint of childlike fantasy to their creation, but this is no product of whimsy; it is the amazing result of equal parts talent, inspiration, hard work, and perhaps a touch of impish delight.

Above: The stair hall from the dining room. Opposite, bottom left and center: Dining room. Opposite, top: A bedroom in the uppermost reaches of the house has views of the surrounding property and is decorated with wall designs crafted by Randy Korando from pine cones, nuts, acorns, and vines. Opposite, far right: The turreted porch was formed by joining two half-round choir lofts.

CAMP BOXWOODS

The grounds of Camp Boxwoods encourage lingering and looking, with views through gates and hedges, over lakes, and under hanging limbs. Above and opposite: The stables were configured from an abandoned church to minister to rescued and rehabilitated horses.

McFaddin Townhouse
110 Tuell Court
1918; renovated in 1986

Ginger McFaddin is one of the earliest pioneers of this newly redeveloping downtown area in the Madison Historic District. She bought the property in 1985 and spent two years renovating and adapting the old commercial building as a residence. She says the results realized "95% of my first vision for the place."

A native of New York State, with a love of cats and hats, Ginger moved here from Atlanta to operate a consulting business that had outgrown her home. Trusting her instincts, she proceeded to carve out an inviting living space in a spot some local citizens would have considered most unlikely. Entered from a landscaped alley courtyard, a wall-free, twenty-five-foot atrium reflects a cosmopolitan lifestyle that is becoming more prevalent in historic Madison than it was when Ginger moved here more than twenty years ago. All the spaces merge and converge on a three-sided island kitchen illuminated by brass industrial lamps hung from the ceiling. A spiral

Opposite, top: The entrance courtyard is a haven just off a busy downtown street. Opposite, bottom: The entrance door is inset and suspended in an aperture of glass. Above: The living room and kitchen are open to loft space accessed by a spiral stair.

staircase leads to a balcony loft, with pressed metal ceilings throughout. *The Madisonian, Architecture and History* (1992) describes this townhouse as "distinctive downtown architecture," fashioned from a 1918 brick-and-mortar business building.

Conceived as a Packard dealership, the building was forced into an immediate adaptive-use conversion by the boll weevil, as the local cotton economy declined and the demand for new cars fell with it. At times it has sheltered mules and cars and cotton, but now it is home to an ever-changing assortment of Maine Coon cats, bred and raised by Ginger and her veterinarian husband, Sammy McFaddin.

HUDSON LOFT
150 WEST JEFFERSON
c. 1885; RENOVATED IN 2001

"Live/work is the idea," Gary Hudson said, his sentiment being the foundation of the standard shopkeeper tradition in urban living and one making a significant comeback in Madison. Gary is an abstract expressionist artist, his wife Christy is a couturier, and this is where they live and work. They moved to this Madison downtown redevelopment area, just west of Main Street, in 2001 from Jefferson, Georgia. Here they have two floors, six thousand square feet. Although not exactly fitting the normal shopkeeper mold in downtown residential living, the ground floor does house their work areas, and upstairs are their residence and a gallery for Gary's paintings. Their building dates from around 1885 and was the home for a time of Farmers Hardware Company. Still doing business on South Main Street, Farmers Hardware was founded before the Civil War and is one of the first such stores in the South.

Opposite: The Hudsons access their shop, studio, and loft from the side street. Opposite, bottom: A three-dimensional logo marks the entrance to Christy Hudson's shop. Above: Gary Hudson's studio has access to the loft above by a freight elevator. Below: Dining area.

Gary Hudson is a leading member of the abstract expressionist school, a post–World War II American art movement centered in New York City in the 1950s and '60s. Among his colleagues were such figures as Franz Kline, Clyfford Still, Willem de Kooning, and Helen Frankenthaler. For such a modern era artist, Hudson maintains a real interest in American history, especially the Revolutionary War period.

Hudson's large, colorful canvases require a work space where he can spread out, as does Christy Hudson's custom-made dressmaking operation. His work was featured in Atlanta in the spring of 2009 at the Forward Arts Foundation's Swan Coach House Gallery at the Atlanta History Center. Gary and Christy find downtown Madison to be a stimulating environment in which to live and work.

The living areas are designed to display art and provide comfortable space for gathering with friends. Frame timbers, open rafters, and bare brick are reminders of the building's origins. Above and opposite, lower left: The large living room is a gallery and center for conversation. Opposite, top: Family room and kitchen. Opposite, lower right: Mosaic above the kitchen sink.

HUDSON LOFT

Long Residence
271 West Washington Street
c. 1910; renovated in 2004

Above: The Longs' balcony looks out over the new Town Park. Opposite: The living room, dining room, den, office, and kitchen spaces are defined by groupings of furniture, partitions, and a loft. Stickley pieces, Cherokee artifacts, and structural timbers are distinctive design elements.

Paulette and Lloyd Long moved into this urban loft space in 2004, the same year construction began on Town Park, the multimillion-dollar redevelopment project across the street, two blocks west of the town square. That was early in the renewal of this West Jefferson and West Washington street section of the downtown Madison Historic District, making the Longs certifiable residential pioneers in this former warehouse and commercial district. Built circa 1910, this brick building known as the Ice House had originally been the Madison Fertilizer Company and later the Farmers Trading Company. By November 30, 2006, their Ice House loft was included in the Heritage Holiday Tour of Homes, along with more conventional tour sites such as Thurleston, Hilltop, and the Rogers House Museum.

The Longs had moved from the 1917 vernacular Prairie style Douglas-McDowell house on South Main Street, itself a modern era landmark, where their Arts and Crafts furnishings had also been very much at home. Their interior decorator was Madisonian Barbara (Bobbie) Rushing, who also designed the interiors of the James Madison Inn.

In the guide to the 2006 Heritage Tour of Homes, Paulette Long acknowledged Rushing: "Thanks to our wonderful interior designer, Bobbie Rushing, every place in our home fits us perfectly." Paulette said her "home's most distinguished feature [was] Lloyd's wonderful collection of Cherokee Artifacts and fantastic art by our friend Gary Hudson." Lloyd is of Cherokee Indian descent, and Gary's live/work loft is also in this portfolio.

With the Ice House so close to the railroad tracks, the Longs relate that whenever a train chugs through this downtown neighborhood, their grandchildren say, "Grandma, there is your train!"

LONG RESIDENCE

Royal Penthouse
262 West Washington Street
2008

Above: The James Madison Inn. The Royals' interiors blend early twentieth-century architectural styles with a cool palette of wall and fabric colors to create a relaxing atmosphere. Below: Kitchen. Opposite, top: Living room. Opposite, bottom: Dining room.

A penthouse in the heart of the historic district could seem contradictory in a town as small as Madison, but not today. Everett and Jane Royal, former Atlantans (Everett a native, Jane originally from Long Island), have led in the development of the Town Park area west of Main Street, in particular their James Madison Inn and Madison Markets. The city razed a block of non-historic commercial and industrial structures in 2004 to begin construction of Town Park, following the earlier development of nearby Round Bowl Spring Park. The proximity of these parks to their property and the initiative of the City of Madison to support downtown revitalization encouraged the Royals to go forward with an ambitious adaptive use of old warehouses and the construction of new buildings. The Royals' commitment to the town center is helping build new traditions in Madison's modern era.

In its large-scale nod to neoclassicism and themed interiors, the James Madison Inn pays homage to Madison's past. Each of nineteen rooms is named for a landmark Madison home with portraits by local artists of that home above each man-

tel. The Royals' penthouse, completed in 2008 atop the inn, is a study in sophisticated classicism and contentment, which can be attributed to Jane Royal's cousin, Tricia Foley, a nationally known interior designer and author from New York, who worked closely with the Royals to interpret their lifestyle and tastes.

From the penthouse one can take a panoramic visual tour over the trees and rooftops to monuments rising above the Madison townscape, each representing an important part of the small-town story—Godfrey's grain mill, Golucke's Beaux Arts courthouse dome, and the steeples of the historic churches. On the hill beyond Round Bowl Spring is the Madison Old Cemetery, a peaceful and ever-present reminder that history, in this case two hundred years of it, is shaped and told by real people.

Epilogue

Above: The Boat House (c. 1850; enlarged c. 1900 and 1984). Opposite: The porch of the Owen-Landry house looks out on Academy Street.

Writing a commemorative history of a two-hundred-year-old town may or may not turn up any new insights, interpretations, or facts. The endeavor usually begins and ends with the well-known traditions, stories, and personalities of the place, and research through traditional sources sometimes does little more than repeat popular tales.

In the historical essay we dealt with the time-worn issue of Madison's signature old story—that the town was spared because it was too beautiful for General Sherman to burn. Let us close the book by celebrating another time-honored phrase, one coined by a French historical philosopher named Jean-Baptiste Alphonse Karr (1808–90) in 1849: *Plus ça change, plus c'est la même chose.* (The more things change, the more they remain the same.)

Things obviously change in Madison, but through conscious community effort, the best characteristics of the town have remained very much the same. There has long been an emphasis on historic preservation, on enhancing the inherited historical character of one of the state's largest historic districts. The Morgan County Cultural Center, a preserved and adapted 1895 grade school in the very heart of things, symbolizes and fosters the town's history and heritage—its educated culture. Madison has remained classic, yet lively, a place with a recognizable Main Street way of life. Madison is a destination to experience, to share in, and, finally, often to settle into, as people have been doing since the original settlers in 1809. And, Dear Reader, in this book we have tried to tell you the truth, the continuing truth, about two-hundred-year-old Madison, a place that changes only enough year by year so that each generation will love it as always and anew.

EPILOGUE

Selected Sources and Observations

Books and Other Publications

Banks, William. "History in Towns, Madison, Georgia." *The Magazine Antiques*. (April 2009) 88–97.

Carter, Christine Jacobson, ed. *The Diary of Dolly Lunt Burge, 1848–1879*. Athens: University of Georgia Press, 1997. Burge lived in Madison and the area. This is the latest edition of the well-known diary.

Chapman, William. *The Madison Historic Preservation Manual, a Handbook*: The Madison Historic Preservation Commission and the City of Madison, Georgia, 1990. This is an excellent history and historic preservation resource.

Cobb, James C. *Away Down South: A History of Southern Identity*. Oxford; New York: Oxford University Press, 2005.

——. *Georgia Odyssey*. 2nd ed. Athens: University of Georgia Press, 2008. Published in association with the Georgia Humanities Council.

Cook, Jacquelyn. *The Gates of Trevalyan*. Smyrna, Georgia: Belle Books, 2008. A novel set in antebellum Madison/Morgan, it features actual historical settings and persons, such as Joshua Hill and Alexander H. Stephens.

Flanders, Ralph Betts. *Plantation Slavery in Georgia*. Cos Cob, Conn.: J.E. Edwards, 1967.

Green, Henry D. *Furniture of the Georgia Piedmont before 1830*. Atlanta: The High Museum of Art, 1976.

Hicky, Louise McHenry. *Rambles through Morgan County*. Ed. Morgan County Historical Society. Madison, 1971. This is a fine published resource for local lore.

Hitchcock, Susan L. "Teacher's Heritage Resource Guide." Georgia Trust for Historic Preservation, Inc., and Morgan County Landmarks Society, 1999.

Jones, Charles C., Jr. *The History of Georgia*. Boston: Houghton, Mifflin and Co., 1883.

Jones, Charles Edgeworth. *Education in Georgia*. Washington: Government Printing Office, 1889.

Jones, Joseph. *Major Jones's Courtship*. Edited by William Tappan Thompson. Philadelphia: T. B. Peterson, 1843.

Kaemmerlen, Cathy. *General Sherman and the Georgia Belles: Tales from Women Left Behind*. Charleston, S.C.: History Press, 2006. Includes Dolly Lunt Burge and Emma High of Madison, Georgia.

Kahn, E. J. *Georgia from Rabun Gap to Tybee Light*. Atlanta: Cherokee Publishing Co., 1978. Published as a two-part series in the *New Yorker* magazine in 1977, it has an interview with the Hicky family in their Madison home.

Kay, Terry. *To Dance with the White Dog: A Novel*. Atlanta: Peachtree Publishers, 1990.

Kennett, Lee B. *Sherman: A Soldier's Life*, 1st ed. New York: HarperCollins, 2001. His discussion of Sherman's Special Field Order Number 120 is detailed.

Knight, Lucian Lamar. *A Standard History of Georgia and Georgians*. Chicago and New York: Lewis Publishing Co., 1917.

——. *Georgia's Landmarks, Memorials, and Legends*, vol. 3. Atlanta: Byrd Printing Co., 1913/14.

Lane, Mills, ed. *"War Is Hell!" William T. Sherman's Personal Narrative of His March through Georgia*. Savannah: Beehive Press, 1974.

Linley, John. *Architecture of Middle Georgia: The Oconee Area*. Athens: University of Georgia Press, 1972.

——, and Historic American Buildings Survey. *The Georgia Catalog, Historic American Buildings Survey: A Guide to the Architecture of the State*. Athens: University of Georgia Press, 1982. This includes valuable architectural references to Madison/Morgan.

Madison, Georgia, an Architectural Guide. Madison: Madison-Morgan Cultural Center, 1991.

Martin, Van Jones, and William R. Mitchell Jr. *Landmark Homes of Georgia, 1733–1983: Two Hundred and Fifty Years of Architecture, Interiors, and Gardens*. Savannah: Golden Coast Publishing Co.,1982. Pages 102–11 cover aspects of Madison/Morgan.

McAlester, Virginia and Lee. *A Field Guide to American Houses*. New York: Knopf, 1994. On page 190 is 611 N. Main Street, Madison, the Massey-Harris house, given as an example of the Greek Revival style.

McCrady, Allston. "Boxwood." *Garden and Gun*. Summer 2007, 80–86.

Melton, Brian. "The Town That Sherman Wouldn't Burn, Sherman's March and Madison, Georgia, in History, Memory, and Legend." *The Georgia Historical Quarterly*. Summer 2002, 201–30. This article is a good source on Joshua Hill, Madison, and Sherman's March to the Sea.

Mitchell, William R., Jr., and Richard Moore. *Gardens of Georgia*. Atlanta: Peachtree Publishers, with the Garden Club of Georgia, 1989.

Morgan County, Georgia Heritage, 1807–1992. Morgan County Historical Society.

"Morgan County, Georgia, History and Biographies." Reprint, 2001.

Myers, Theodorus Bailey, ed. *Being Correspondence of General Morgan and the Prominent Actors*. Charleston, S.C.: The *News and Courier*, 1881.

Nesbitt, R. J. *Georgia: Her Resources and Possibilities*. Atlanta: Franklin Printing Company, 1895. This is just one of several such geographies, arranged county by county, that include Madison/Morgan. On page 452 is this statement: "the county seat, Madison, is one of the most beautiful towns in Georgia."

"New Directions in Preservation: Georgians Research Georgia." In *Sixth Annual Conference on Historic Preservation May 23–25, 1974*. Atlanta: Georgia Trust for Historic Preservation, 1974/75. This is the author's own copy of this rare imprint. It records his participation in the conference as a moderator on Georgia architecture.

Newton, Floyd C., Jr. *Directory of Graves, Old Madison Cemetery*. Madison, Georgia: Bank of Madison, 2001.

"Oak House, Absolute Auction Package, the Dupree Residence." 2006.

"Opening a Window to Morgan County's African-American Heritage." In *Special Supplement to the Madisonian*. Madison, Georgia: Madison County African-American Museum, 1993.

Perkerson, Medora Field. *White Columns in Georgia*. New York: Rinehart, 1952. A well-researched popular account. Perkerson includes, on page 46, the well-known quotation from a Yankee soldier in November 1864 that describes Madison as "the prettiest village I've seen in the state."

Ponder, Adelaide W., ed. *The Madisonian, Architecture and History*. W. Graham Ponder, 1992. The Ponders were well-qualified Morgan County natives who published several editions of the valuable folio for their newspaper.

Poss, Faye Stone. *The Southern Watchman*, 1861–1865. The Watchmen was published in Athens, Georgia.

Pratt, Dorothy, and Richard. *A Guide to Early American Homes: North and South*. New York: Bonanza Books, 1956.

Rodgers, Ava D. *The Housing of Oglethorpe County, Georgia, 1790–1860*. Tallahassee: Florida State University Press, 1971. One of the best sources of architectural data on the Plantation Plain style, "I-house," folk vernacular.

Savor Madison: A Culinary Tour. Church of the Advent Episcopal Church. Madison, Georgia, 1989.

Scruggs, C. P. *Georgia Historical Markers:* Complete Texts of 1752 Markers. Valdosta: Bay Tree Grove Publishers, 1973. There are twelve of the Madison/Morgan state historical markers. Markers with data about the old roads that passed through the county and served emigrants heading west are especially helpful, as are those about the March to the Sea.

Spector, Tom, and Susan Owings-Spector. *The Guide to the Architecture of Georgia*. Columbia, S.C.: University of South Carolina Press, 1993. This little guidebook includes a good section on Madison/Morgan, with a map.

Trudeau, Noah Andre. *Southern Storm: Sherman's March to the Sea*. New York: Harper, 2008. This is an excellent new book, with numerous original references to Madison. On page 140, Trudeau quotes an Illinois Yankee who described the place in 1864 as the "finest village this side of Nashville."

Vaughn, Ralph E., ed. *Madison-Morgan County: 2006–2007 Guide*. Madison: Madison-Morgan Chamber of Commerce, 2007.

Weeks, Christopher, ed. *Palladio and America*. New Orleans: Martin-St. Martin Publishing Company, 1997.

White, George. *Statistics of the State of Georgia*. Savannah: W. Thomas Williams, 1849.

Williams, Clint. "Madison, Small Town, Big Changes." *Atlanta Journal-Constitution*, March 26, 2008, E1 and E7.

Winchester, Alice., ed. *Living with Antiques: a Treasury of Private Homes in America*. New York,: E.P. Dutton, 1963. Henry Green's Greenoaks Plantation is featured on pages 144–51, "Georgia Plantation."

Williams, Philip Lee. *Elegies for the Water*. Macon, Georgia: Mercer University Press, 2009.

Zelinsky, Wilbur. "The Greek Revival House in Georgia." *Journal of the Society of Architectural Historians*. 13. May 1954: 9–12.

Unpublished

National Register of Historic Places (a Federal/State program) nomination forms for Morgan County and Madison. These forms have detailed data. There are districts and individual sites. Lists can be accessed online. The nomination for Bonar Hall, added to the register in 1972, was prepared by the author, as was the form for Cedar Lane Farm, 1971.

P. Thornton Marye, an architect. Original drawings for "Boxwood," prepared for *The Garden History of Georgia, 1733–1933*, are on file at the Georgia Archives. Also see *Garden and Gun,* Summer 2007, 80–81.

Visual Sources

There are numerous archival images, from various sources. Also, many movies were set in Madison. Among them: in 1977, *The Great Bank Hoax* and in 1978, *The Summer of My German Soldier*.

Internet Sources:

georgiainfo.galileo.usg.edu. *Treaty of Washington*.
nps.gov. *The Battle of Cowpens*, by Scott Withrow.
files.usgwarchives.net. *Old Madison Cemetery*.
doughboysearcher.tripod.com. *The E. M. Viquesney "Spirit of the American Doughboy" Database*.
genforum.genealogy.com. *Shaws/Newton-Morgan Co's., GA>1772–*
newgeorgiaencyclopedia.org. *Raymond Andrews,* by Philip Lee Williams.

Interviews

Monica Callahan
Rick Crown
Dog Ear Books
Charles R. Davis
Bonnie P. (Patsy) Harris
Julie Green Jenkins
Annie Jones
Barry Lurey
Fred Perriman

Adelaide Ponder
Paul Reid
Steve Schaefer
Joe Smith
Jane Campbell Symmes
Ken Thomas
Frank Walsh
Marshall "Woody" Williams
Philip Lee Williams

Index of Sites Illustrated

Albert Douglas house, 50
Atkinson house, 49, 148
Atkinson-Rhodes house, 154–57
Baldwin-Williford-Ruffin house, 62, 116
Barnett-Stokes house, 94–97
Billups-Tuell house, 46, 122–25
Boat House, 235
Bonar Hall, 44, 51, 102–3
    (cottage moved from South Main street to Bonar Hall, 65)
Bookhaven, 192–93
Boxwood, 21, 47, 67, 140–47
Broughton-Sanders house, 67
Burney-Ponder-Rushing house, 198–201
Calvary Baptist Church, 29, 50
Camp Boxwoods, 191, 216–23
Candler Lane, 190
Carter-McManus house, 194–95
Carter-Newton house, 46, 64
CCC camp at Rutledge, 57
Cedar Lane Farm, viii, 78–81
Church of the Advent (former Methodist Church), 21, 27, 45, 239
Clark's Chapel Baptist Church, 28, 49
Cooke house, 66
Confederate Monument in Hill Park, 238
Cornelius Vason house, 55
Douglas-McDowell house, 55, 196–97
Edmund Walker Town House, 70–73
Eighth District Agricultural and Mechanical School, 24, 54
First United Methodist Church, 55
Foster house, 150
Georgia Female College (Baldwin-Williford-Ruffin house), 46
Gilbert house, 202–5
Godfrey-Hunt house, 56, 172–75
Henry Lane dogtrot log cabin, 65
Heritage Hall, 54, 104–7
Hilltop, 44, 66, 82–87
Honeymoon 46, 112–15
Hudson Loft 226–29
Hunter house 149
James Madison Inn, 61, 191
John Buck Swords house, 35
Joshua Hill house, 53, 150

La Flora, 164–67
Lake Rutledge, part of Hark Labor Creek Park, 34
Long residence, 230–31
Madison Baptist Church, 26, 47
Madison City Hall (1887), 53, 189
Madison City Hall (1939), 57
Madison Graded School, 25, 52
Madison High School, 23
Madison Hotel, 52
Madison-Morgan Cultural Center, 13, 23, 25
Madison Old Cemetery, 31, 33, 42
Madison Post Office, 57, 189
Madison Presbyterian Church, 26–27, 44
Madison Town Park, 61, 191
Madison Water Works, 54
Madisonian building, 32

Magnolia house, 160–63
Main Street, Madison, x, 22
Martin-Baldwin-Weaver house, 46, 108–11
Martin Richter house, 51, 149
Massey-Tipton-Bracewell house, 46, 126–27
McFaddin Townhouse, 224–25
Morgan County African-American Museum, 29, 170–71
Morgan County Archives, 52
Morgan County Courthouse (1845), 19, 45
Morgan County Courthouse (1907), 19, 45, 53, 58, 60, 151
Morgan County High School, A&M building, xi
Morgan County Jail (now Morgan County Archives), 52
Nathan Bennett house, 134–39
Oak house, (Trammell house), 53, 176–181
Owen-Landry house, 148
Park's Mill, 35
Peter Walton house, 52, 149
Porter-Fitzpatrick house, 53, 182–85
Pool house, 188
Poullain Heights 53, 186–187
Reuben Rogers house, 20, 41, 66, 68–69
Richter Cottage, 74–75
Robson-Mason house, 98–101
Rose Cottage, 152–53
Round Bowl Spring Park, 20
Royal Penthouse, 232–33
Saffold house, 66
Samuel Hanson house, 206–11
Saye house, 189
Shaw-Erwin house, 158–59
Snow Hill, 43
South Main Street School, 52
St. Paul African Methodist Episcopal Church, 29, 50
Stagecoach House, 76–77
Stoke Farm, 214–15
Stokes-McHenry house, 128–33
Thurleston, 45, 88–93
Turnell-Butler Hotel, 52
United Methodist Church, 28
Valley Farm, 190
Vason house, 188
White-Lyle cottage, 168–69
Willow Oak Farm, 212–13

Opposite: The Confederate Monument was moved from downtown to Hill Park in 1955. Above: Church of the Advent.